Directed Reading A

AF251773

Section: The Cardiovascular System

YOUR CARDIOVASCULAR SYSTEM

Match the correct description with the correct term. Write the letter in the space provided.

_______ **1.** The heart and blood vessels are part of this system.

_______ **2.** Blood vessels carry this throughout the body.

_______ **3.** Blood is pumped through the body by this.

_______ **4.** The cardiovascular system helps maintain this.

a. blood

b. cardiovascular

c. homeostasis

d. heart

THE HEART

Write the letter of the correct answer in the space provided.

_______ **5.** The heart is about the size of which of the following?
   a. your head
   b. your nose
   c. your thumb
   d. your fist

_______ **6.** What are the heart's upper chambers called?
   a. atria
   b. ventricles
   c. valves
   d. cardios

_______ **7.** What are the heart's lower chambers called?
   a. atria
   b. ventricles
   c. valves
   d. cardios

_______ **8.** What kind of blood gets sent to the lungs?
   a. type A
   b. type B
   c. oxygen-rich
   d. oxygen-poor

Directed Reading A *continued*

_______ **9.** What kind of blood gets sent to the body?
    a. type A
    b. oxygen-rich
    c. type B
    d. oxygen-poor

_______ **10.** What causes the sound of a heartbeat?
    a. atria contracting
    b. valves closing
    c. ventricles contracting
    d. atria relaxing

BLOOD VESSELS

Match the correct description with the correct term. Write the letter in the space provided.

_______ **11.** carry blood away from the heart

_______ **12.** allow exchanges between blood and cells

_______ **13.** carry blood to the heart

_______ **14.** caused by rhythmic contractions of the heart

_______ **15.** includes arteries, capillaries, and veins

a. capillaries
b. arteries
c. blood vessels
d. veins
e. pulse

TWO TYPES OF CIRCULATION

Match the correct description with the correct term. Write the letter in the space provided.

_______ **16.** flow of blood between heart and lungs

_______ **17.** flow of blood between heart and the rest of the body

a. systemic circulation
b. pulmonary circulation

CARDIOVASCULAR PROBLEMS

Write the letter of the correct answer in the space provided.

_______ **18.** What can lower the risk of cardiovascular problems?
    a. smoking
    b. eating a healthy diet and exercising
    c. having high levels of cholesterol
    d. avoiding exercise

Directed Reading A *continued*

Atherosclerosis

_______ **19.** What is cholesterol buildup in an artery called?
  a. hypertension
  b. heart attack
  c. heart failure
  d. atherosclerosis

_______ **20.** What can cause a narrowing of the arteries?
  a. stroke
  b. heart attack
  c. heart failure
  d. atherosclerosis

High Blood Pressure

_______ **21.** What is another name for high blood pressure?
  a. hypertension
  b. heart attack
  c. heart failure
  d. atherosclerosis

_______ **22.** What can happen when a brain artery clogs?
  a. heart failure
  b. heart attack
  c. stroke
  d. atherosclerosis

Heart Attacks and Heart Failure

_______ **23.** What can happen when the heart muscle does not get enough blood?
  a. hypertension
  b. heart attack
  c. heart failure
  d. atherosclerosis

_______ **24.** What happens when the heart cannot pump enough blood?
  a. hypertension
  b. heart attack
  c. heart failure
  d. atherosclerosis

Directed Reading A

Section: Blood

Write the letter of the correct answer in the space provided.

_______ **1.** How much blood does an adult have?
   a. 5 liters
   b. 10 liters
   c. 50 liters
   d. 100 liters

COMPONENTS OF BLOOD

_______ **2.** What system is made of the heart, blood vessels, and blood?
   a. skeletal system
   b. muscular system
   c. digestive system
   d. cardiovascular system

_______ **3.** What is blood made of?
   a. oxygen and plasma
   b. red blood cells and white blood cells
   c. plasma, red blood cells, platelets, and white blood cells
   d. plasma and platelets

Plasma

_______ **4.** What is plasma?
   a. only white blood cells
   b. only red blood cells
   c. fluid part of blood
   d. hemoglobin

Red Blood Cells

_______ **5.** What are the most common blood cells?
   a. red blood cells
   b. white blood cells
   c. platelets
   d. plasma

_______ **6.** Which cells receive oxygen from red blood cells?
   a. all cells
   b. only skin cells
   c. only muscle cells
   d. only bone cells

_______ **7.** What attaches to the oxygen you breathe and carries oxygen on red
blood cells?
a. plasma
b. hemoglobin
c. platelets
d. bone marrow

Platelets

_______ **8.** Where are platelets made?
a. plasma
b. bone marrow
c. white blood cells
d. red blood cells

_______ **9.** Why do platelets clump together?
a. to produce oxygen
b. to reduce oxygen
c. to produce blood loss
d. to reduce blood loss

White Blood Cells

_______ **10.** What are pathogens?
a. disease-causing bacteria, viruses, and other microorganisms
b. large platelets
c. antibodies
d. tiny fibers

_______ **11.** What destroys pathogens?
a. red blood cells
b. white blood cells
c. platelets
d. plasma

_______ **12.** What part of the blood destroys dead and damaged cells?
a. white blood cells
b. red blood cells
c. platelets
d. pathogens

BODY TEMPERATURE REGULATION

Match the correct description with the correct term. Write the letter in the space provided.

_______ **13.** helps regulate your body temperature

_______ **14.** enlarge when your body temperature rises

_______ **15.** lowers when heat is transferred from blood to skin

a. blood

b. temperature

c. blood vessels

BLOOD PRESSURE

Match the correct description with the correct term. Write the letter in the space provided.

_______ **16.** force of blood pushing on walls of arteries

_______ **17.** pressure inside large arteries when ventricles contract

_______ **18.** pressure inside arteries when ventricles relax

a. diastolic

b. systolic

c. blood pressure

BLOOD TYPES

Match the correct description with the correct term. Write the letter in the space provided.

_______ **19.** chemicals on red blood cells that determine blood type

_______ **20.** antigens in type A blood

_______ **21.** antigens in type B blood

a. antigens

b. A antigens

c. B antigens

TRANSFUSIONS AND BLOOD TYPES

Write the letter of the correct answer in the space provided.

_______ **22.** What does a transfusion replace?
    a. lost body temperature
    b. lost pathogens
    c. lost blood
    d. lost antibodies

_______ **23.** What could happen if you receive the wrong blood type?
    a. Your blood type could change.
    b. You might need more white blood cells.
    c. You might get too much oxygen.
    d. You could die.

Directed Reading A *continued*

BLOOD DISORDERS

Match the correct description with the correct term. Write the letter in the space provided.

_______ **24.** condition in which blood does not clot normally

_______ **25.** cancer that affects blood cells

a. leukemia

b. hemophelia

Directed Reading A

Section: The Respiratory System

Write the letter of the correct answer in the space provided.

_______ **1.** Why does the body need oxygen?
    a. to get energy from food
    b. to make its own food
    c. to fight infection
    d. to make more blood

RESPIRATION AND THE RESPIRATORY SYSTEM

Match the correct description with the correct term. Write the letter in the space provided.

_______ **2.** the process of using oxygen and releasing carbon dioxide and water

_______ **3.** the process of inhaling and exhaling

_______ **4.** the organs that take in oxygen and get rid of carbon dioxide

    a. respiration
    b. respiratory system
    c. breathing

Nose, Pharynx, and Larynx

_______ **5.** the main passage into and out of the respiratory system

_______ **6.** the part of the throat that produces sounds

_______ **7.** the throat

    a. pharynx
    b. nose
    c. larynx

Trachea

Write the letter of the correct answer in the space provided.

_______ **8.** What is the trachea also called?
    a. nose
    b. throat
    c. tonsils
    d. windpipe

_______ **9.** What goes through the trachea?
    a. blood to the heart
    b. air to the lungs
    c. food to the stomach
    d. lymph to the lymph nodes

Directed Reading A *continued*

Bronchi and Alveoli

_______ **10.** What is a tube connecting the lungs with the trachea?
    a. pharynx
    b. nose
    c. larynx
    d. bronchus

_______ **11.** What are bronchioles?
    a. smaller branches of bronchi
    b. tiny sacs in the lungs
    c. tubes next to the larynx
    d. pharynx

_______ **12.** What are alveoli?
    a. smaller branches of bronchi
    b. tiny air sacs in the lungs
    c. tubes next to the larynx
    d. pharynx

BREATHING

Match the correct description with the correct term. Write the letter in the space provided.

_______ **13.** contracts and moves down when you inhale   **a.** diaphragm

_______ **14.** contract and lift the rib cage   **b.** rib muscles

Breathing and Cellular Respiration

Match the correct description with the correct term. Write the letter in the space provided.

_______ **15.** When you inhale, you take in this.   **a.** energy

_______ **16.** Cells use oxygen to release this.   **b.** oxygen

Directed Reading A *continued*

Respiratory Disorders

Write the letter of the correct answer in the space provided.

_______ **17.** What may trigger asthma?
- **a.** blood cells
- **b.** dust or pollen
- **c.** antigens
- **d.** SARS

_______ **18.** What causes SARS?
- **a.** blood cells
- **b.** dust or pollen
- **c.** virus
- **d.** bacteria

_______ **19.** Which of the following might people with respiratory disorders have trouble with?
- **a.** getting rid of oxygen
- **b.** gaining carbon dioxide
- **c.** getting rid of carbon dioxide
- **d.** gaining too much energy

Directed Reading B

Section: The Cardiovascular System

1. The word *cardio* means _____________________.

2. The word *vascular* means _____________________.

3. Arteries, veins, and capillaries are _____________________ that carry blood pumped by the heart.

YOUR CARDIOVASCULAR SYSTEM

4. The heart, blood, and blood vessels together make up

the _____________________.

5. What are three examples of how the cardiovascular system helps maintain homeostasis?

____

____

____

THE HEART

______ **6.** What kind of tissue makes up most of the heart?
    a. vascular   **c.** circulatory
    b. cardiac   **d.** lymphatic

______ **7.** Each upper chamber of the heart is called a(n)
    a. atrium.
    b. ventricle.
    c. valve.
    d. cardio.

______ **8.** Each lower chamber of the heart is called a(n)
    a. atrium.
    b. ventricle.
    c. valve.
    d. cardio.

______ **9.** The right side of the heart pumps oxygen-poor blood to the
    a. body.
    b. lungs.
    c. right ventricle.
    d. left atrium.

Name _________________________________ Class _________________ Date _______________

______ **10.** The left side of the heart pumps oxygen-rich blood to the
      a. body.
      b. lungs.
      c. right ventricle.
      d. left atrium.

______ **11.** When atria relax, what do ventricles do?
      a. expand
      b. relax
      c. contract
      d. close

12. Why do the heart's valves close?

____

____

13. What part of the heart makes the heartbeat sound?

____

BLOOD VESSELS

Use the terms from the following list to complete the sentences below.

    capillaries   arteries   blood vessels
    veins   pulse

14. Blood vessels that carry blood away from the heart are

_________________________.

15. Blood vessels that allow exchanges between blood and cells are

_________________________.

16. Blood vessels that carry blood to the heart are _______________________.

17. Your _____________________ is caused by rhythmic

changes in your blood pressure.

18. Blood travels through your body in hollow tubes called

_________________________.

Directed Reading B *continued*

19. What is the narrowest kind of blood vessel?

____

20. What kind of blood vessels are helped by the contracting of skeletal muscles?

____

21. What kind of blood vessels stretch due to blood pressure?

____

TWO TYPES OF CIRCULATION

_______ **22.** What enters the blood when it is pumped to the lungs?
  a. blood vessels
  b. capillaries
  c. oxygen
  d. carbon dioxide

_______ **23.** What leaves the blood when it is pumped to the lungs?
  a. blood vessels
  b. capillaries
  c. oxygen
  d. carbon dioxide

_______ **24.** Where does the exchange of blood and oxygen take place in the lungs?
  a. ventricles
  b. arteries
  c. capillaries
  d. veins

25. The flow of blood between the heart and the lungs is

______________________ circulation.

26. The flow of blood between the heart and the rest of the body

is ______________________ circulation.

CARDIOVASCULAR PROBLEMS

27. Cardiovascular problems can harm the whole ______________________.

28. A healthy diet and plenty of ______________________ can reduce the risk
of cardiovascular problems.

29. What is the leading cause of death in the United States?

____

____

Directed Reading B *continued*

Match the correct description with the correct term. Write the letter in the space provided. Some terms will not be used.

_______ **30.** fatty buildup in blood vessels

_______ **31.** abnormally high blood pressure

_______ **32.** brain damage caused by damage to blood vessels

_______ **33.** condition caused by the death of heart muscle cells

a. blood poisoning

b. heart failure

c. heart attack

d. atherosclerosis

e. stroke

f. hypertension

34. What can happen when blood supply to the heart is blocked?

____

____

____

Directed Reading B

Section: Blood

1. An adult body has about _________________________ liters of blood.

COMPONENTS OF BLOOD

2. List the four components of blood.

____

____

____

____

3. The fluid part of blood is called _________________________.

4. What do red blood cells do?

____

____

5. To what does hemoglobin attach?

____

6. When bleeding starts, _________________________ form a plug to reduce

blood loss.

7. What two things do white blood cells do?

____

____

8. Pathogens include bacteria, _________________________, and other

microscopic organisms.

9. To fight pathogens, white blood cells destroy pathogens or

release _________________________.

10. What destroys dead or damaged body cells?

____

____

BODY TEMPERATURE REGULATION

11. Blood helps to regulate body _____________________.

12. When body temperature rises, blood vessels in the skin

_____________________.

BLOOD PRESSURE

______ **13.** The force that blood exerts on the walls of the arteries is called
   a. systolic pressure.
   b. blood pressure.
   c. contraction.
   d. diastolic pressure.

14. When is systolic pressure measured?

____

____

15. When is diastolic pressure measured?

____

____

16. What are two organs that can be damaged by high blood pressure?

____

____

BLOOD TYPES

______ **17.** To what does a blood type refer?
   a. blood pressure levels
   b. antibodies
   c. blood color and consistency
   d. chemicals called antigens on red blood cells

______ **18.** What kind of antibodies does a person with type A blood have?
   a. type A
   b. type B
   c. type O
   d. type A and type B

| Directed Reading B *continued*

---

19. How do type B antibodies react to type B antigens?

____

____

20. What can happen if an RH⁻ person receives RH⁺ blood?

____

____

____

TRANSFUSIONS AND BLOOD TYPES

_______ **21.** A person with which blood type can donate blood to people of any
other blood type?
- **a.** A
- **b.** B
- **c.** AB
- **d.** O

22. What does a transfusion attempt to replace?

____

____

____

BLOOD DISORDERS

23. What are two common blood disorders?

____

____

Name _________________________________ Class _______________ Date ____________

Directed Reading B

Section: The Respiratory System

1. Your body needs oxygen in order to get _________________ from the food you eat.

RESPIRATION AND THE RESPIRATORY SYSTEM

Match the correct description with the correct term. Write the letter in the space provided.

_______ 2. area of the throat that contains the vocal chords

_______ 3. group of organs that take in oxygen and get rid of carbon dioxide

_______ 4. main passage into and out of the respiratory system

_______ 5. throat

_______ 6. tube guarded by the larynx

_______ 7. tube that connects the trachea to the lungs

_______ 8. smaller branches of bronchi

_______ 9. tiny sacs at the ends of the bronchioles

a. larynx

b. nose

c. pharynx

d. respiratory system

e. alveoli

f. bronchioles

g. bronchus

h. trachea

Match the correct description with the correct term. Write the letter in the space provided.

_______ 10. process of using oxygen and releasing carbon dioxide and water

_______ 11. inhalation and exhalation

_______ 12. use of oxygen by cells to release energy stored in food

a. breathing

b. respiration

c. cellular respiration

BREATHING

13. List the two kinds of muscle involved with breathing.

____

____

Directed Reading B *continued*

14. When you inhale, the _____________________ contracts.

15. When the diaphragm contracts, the _____________________ contract and lift the rib cage.

BREATHING AND CELLULAR RESPIRATION

16. When you inhale, you take in _____________________.

17. Oxygen diffuses into _____________________.

18. When the oxygen is carried to cells, it is used to release

_____________________.

19. Cellular respiration produces water and _____________________.

RESPIRATORY DISORDERS

_______ **20.** Asthma can be triggered by
    a. blood cells.
    b. dust or pollen.
    c. antigens.
    d. SARS.

_______ **21.** SARS is caused by
    a. dust or pollen.
    b. a virus.
    c. bacteria.
    d. asthma.

22. People with respiratory disorders may have trouble getting rid of

_____________________.

Skills Worksheet

Vocabulary and Section Summary A

The Cardiovascular System

VOCABULARY

In your own words, write a definition of the following terms in the space provided.

1. cardiovascular system

____

____

____

2. artery

____

____

____

3. capillary

____

____

____

4. vein

____

____

____

5. pulmonary circulation

____

____

____

6. systemic circulation

____

____

____

Vocabulary and Section Summary A *continued*

SECTION SUMMARY

Read the following section summary.

- Parts of the cardiovascular system include the heart, three types of blood vessels, and blood.
- Contractions of the heart pump blood throughout the body. Valves ensure that blood flows in only one direction.
- The three types of blood vessels are arteries, veins, and capillaries.
- Oxygen-poor blood flows from the heart through the lungs, where it picks up oxygen. Oxygen-rich blood flows from the heart to the rest of the body.
- Cardiovascular problems include atherosclerosis, hypertension, strokes, heart attacks, and heart failure.

Vocabulary and Section Summary A

Blood

VOCABULARY

In your own words, write a definition of the following terms in the space provided.

1. blood

____

____

____

2. blood pressure

____

____

____

SECTION SUMMARY

Read the following section summary.

- The four main components of blood are plasma, red blood cells, platelets, and white blood cells.

- Blood carries oxygen and nutrients to cells, helps protect against disease, and helps regulate body temperature.

- Blood pressure is the force that blood exerts on the inside walls of arteries. It is often expressed in the unit of millimeters of mercury.

- Every person has one of four ABO blood types.

- Losing blood, mixing blood types, and blood disorders can be fatal.

Vocabulary and Section Summary A

The Respiratory System

VOCABULARY

In your own words, write a definition of the following terms in the space provided.

1. respiration

2. respiratory system

3. pharynx

4. larynx

5. trachea

6. bronchus

7. alveolus

Vocabulary and Section Summary A *continued*

SECTION SUMMARY

Read the following section summary.

- Air enters through the nose or mouth, then travels to the pharynx, larynx, trachea, and bronchi. The bronchi branch into bronchioles, which branch into alveoli.

- Breathing involves lungs, muscles in the rib cage, and the diaphragm.

- Oxygen enters the blood through the alveoli in the lungs. Carbon dioxide leaves the blood and is exhaled.

- Respiratory disorders include asthma, emphysema, and SARS.

Skills Worksheet

Vocabulary and Section Summary B

The Cardiovascular System

VOCABULARY

After you finish reading the section, try this puzzle! Use the clues below to fill in the correct terms. Then, complete the word search puzzle on the next page. Words may appear horizontally, vertically, diagonally, or backward.

______________________ **1.** system that transports materials to and from the body's cells

______________________ **2.** type of blood circulation between the heart and the lungs

______________________ **3.** the smallest blood vessels in the body

______________________ **4.** type of blood circulation between the heart and the rest of the body

______________________ **5.** blood vessels that direct blood away from the heart

______________________ **6.** blood vessels that direct blood toward the heart

▌Vocabulary and Section Summary B *continued*

G	L	Y	M	X	A	T	I	C	I	M	E	T	S	Y	S	A	M
T	I	A	L	V	E	E	L	I	T	O	L	N	M	O	V	L	O
O	B	S	U	R	D	S	H	U	O	C	T	E	F	O	S	X	S
N	E	L	A	O	E	C	P	A	B	C	A	C	L	Q	Y	M	C
S	H	T	G	I	N	T	H	Y	T	A	P	L	I	M	R	P	A
I	I	L	R	O	X	Y	L	N	D	R	X	N	Y	R	A	H	P
L	B	A	R	O	D	D	T	B	I	D	I	T	R	E	N	N	I
S	I	B	L	B	I	P	C	Y	G	I	S	A	O	S	O	O	L
E	B	M	A	R	A	L	R	S	U	O	B	O	T	P	M	G	L
L	M	L	H	R	P	A	H	E	E	V	I	X	A	I	L	E	A
C	A	H	G	J	H	T	U	I	S	A	R	L	P	K	U	S	R
I	W	R	O	R	R	E	M	R	P	S	C	S	D	T	P	U	I
R	X	N	H	R	A	L	E	E	L	C	U	B	P	R	S	I	E
T	C	T	V	S	Y	E	Y	T	E	U	P	R	S	I	E	U	S
T	C	A	N	D	M	T	K	R	E	L	I	E	E	O	S	S	R
E	P	E	M	L	A	S	M	A	N	A	Y	M	L	N	T	E	I
V	E	I	N	S	U	Z	Y	H	T	R	A	Z	H	E	A	S	N

SECTION SUMMARY

Read the following section summary.

- Parts of the cardiovascular system include the heart, three types of blood vessels, and blood.
- Contractions of the heart pump blood throughout the body. Valves ensure that blood flows in only one direction.
- The three types of blood vessels are arteries, veins, and capillaries.
- Oxygen-poor blood flows from the heart through the lungs, where it picks up oxygen. Oxygen-rich blood flows from the heart to the rest of the body.
- Cardiovascular problems include atherosclerosis, hypertension, strokes, heart attacks, and heart failure.

Name _________________________________ Class _______________ Date _____________

Vocabulary and Section Summary B

Blood

VOCABULARY

After you finish reading the section, try this puzzle! Use the clues below to solve the crossword puzzle on the following page.

ACROSS

2. force that blood exerts on the walls of arteries (two words)
5. disease in which blood does not clot normally
7. a cancer that affects blood cells
8. fluid that carries gases, nutrients, and wastes throughout the body
11. type of cells that destroy pathogens (three words)
14. chemicals that identify or destroy pathogens

DOWN

1. pressure inside arteries when ventricles contract
3. bacteria, viruses, and other microscopic particles that can make you sick
4. type of cells that carry oxygen throughout the body
5. an oxygen-carrying protein
6. type of molecule on the surface of red blood cells
9. occurs when a person's cells do not get enough blood
10. pieces of larger cells found in bone marrow
12. injection of blood or blood components into a person
13. pressure inside arteries when ventricles relax

Vocabulary and Section Summary B *continued*

SECTION SUMMARY

Read the following section summary.

- The four main components of blood are plasma, red blood cells, platelets, and white blood cells.
- Blood carries oxygen and nutrients to cells, helps protect against disease, and helps regulate body temperature.
- Blood pressure is the force that blood exerts on the inside walls of arteries. It is often expressed in the unit of millimeters of mercury.
- Every person has one of four ABO blood types.
- Losing blood, mixing blood types, and blood disorders can be fatal.

Skills Worksheet

Vocabulary and Section Summary B

The Respiratory System

VOCABULARY

After you finish reading the section, try this puzzle! Use the clues given to fill in the blanks below. Then, copy the numbered letters into the corresponding squares on the next page to reveal the main function of the respiratory system.

1. This includes breathing and cellular respiration.

___ ___ ___ ___ ___ ___ ___ ___ ___ ___
22   3   18   29   21   25   24   12

2. This is also called the throat.

___ ___ ___ ___ ___ ___ ___
          13   5   28

3. This connects the larynx to the lungs.

___ ___ ___ ___ ___ ___ ___
 1   20   9   11

4. This contains the vocal cords.

___ ___ ___ ___ ___ ___
          25   7

5. These connect the lungs to the trachea.

___ ___ ___ ___ ___ ___ ___
23   10

6. This organ system includes the pharynx, larynx, trachea, and lungs.

___ ___ ___ ___ ___ ___ ___ ___ ___ ___ ___ ___
    19   4   6   8

___ ___ ___ ___ ___ ___
16   30

7. These are tiny air sacs in the lungs.

___ ___ ___ ___ ___ ___ ___
 2   15   14   27   26

Vocabulary and Section Summary B *continued*

8. What is the main function of the respiratory system?

SECTION SUMMARY

Read the following section summary.

- Air enters through the nose or mouth, then travels to the pharynx, larynx, trachea, and bronchi. The bronchi branch into bronchioles, which branch into alveoli.

- Breathing involves lungs, muscles in the rib cage, and the diaphragm.

- Oxygen enters the blood through the alveoli in the lungs. Carbon dioxide leaves the blood and is exhaled.

- Respiratory disorders include asthma, emphysema, and SARS.

Reinforcement

Matchmaker, Matchmaker

Complete this worksheet after you finish reading the section "Blood."
As you learned in this chapter, different blood types have different antigens and antibodies. Antigens are chemicals on the surface of red blood cells. Antibodies are chemicals in the blood's plasma. A person makes antibodies against the antigens that his or her red blood cells do not have. Those antibodies will attack any red blood cell that has those antigens, causing the red blood cells to clump together.

BLOOD TYPE	ANTIGENS	ANTIBODIES
O	none	AB
A	A	B
B	B	A
AB	AB	none

A person can receive blood from another person if the donor's blood does not contain antigens that the antibodies of the receiver's blood will attack. Complete the table below by writing *yes* or *no* in each of the blanks.

Receiver's Blood type	Can receive O?	Can receive A?	Can receive B?	Can receive AB?
O				
A				
B				
AB				

Critical Thinking

Doctor for a Day

GRAND AVENUE CLINIC
PATIENT HISTORY FORM

Name: Parker Smith   **Age:** 20   **Doctor:** Anne Williams

Symptoms: unpredictable severe pain in different body tissues, paleness, shortness of breath

Med. History: hospitalized for a stroke at age 7, blood clots detected, frequent blood transfusions

Diagnosis: sickle cell disease

Patients with sickle cell disease have crescent-shaped red blood cells that tend to be sticky and rigid. Sickle cells contain abnormal hemoglobin molecules. Help Dr. Williams explore this disease in more detail.

MAKING COMPARISONS

1. Compare sickle cells with normal red blood cells.

____

____

____

____

SEEING RELATIONSHIPS

2. How could the structure of the sickle cells be related to pain in body tissues?

HELPFUL HINT

Severe pain in body tissues could be caused by a lack of oxygen.

____

____

____

____

| Critical Thinking *continued*

DEMONSTRATING REASONED JUDGMENT

3. Why has Mr. Smith needed frequent blood transfusions?

HELPFUL HINT

When red blood cells are squeezed through the spleen, the more fragile cells break open.

DRAWING CONCLUSIONS

4. Do you agree that Mr. Smith has sickle cell disease? Explain.

MAKING COMPARISONS

5. Would a successful bone marrow transplant be a better long-term treatment than blood transfusions? Explain.

Name _______________________________ Class _____________ Date __________

SciLinks Activity

BLOOD DONATION

Go to www.scilinks.org. To find links related to blood donations, type in the keyword HY70178. Then, use the links to answer the following questions about blood donation.

1. Name and describe two blood components processed from a single blood donation.

   ____

   ____

   ____

   ____

   ____

2. What is one type of test performed on donated blood? Why is this test important?

   ____

   ____

   ____

   ____

3. Define autologous and allogeneic transfusions. Describe a situation in which one or the other might be used.

   ____

   ____

   ____

   ____

   ____

Skills Worksheet

Section Review

The Cardiovascular System

UNDERSTANDING CONCEPTS

1. **Modeling** Describe the pathway of blood flow. Begin and end in the left atrium.

2. **Describing** Describe the functions of the five parts of the cardiovascular system.

3. **Comparing** Compare a heart attack and heart failure.

4. **Analyzing** What is the function of valves?

CRITICAL THINKING

5. **Identifying Relationships** How is the structure of capillaries related to their function?

Section Review *continued*

6. Making Inferences One of aspirin's effects is that it prevents substances in blood from being too "sticky." Why might doctors prescribe aspirin for patients who have had a heart attack?

____

____

____

____

7. Analyzing Ideas Veins and arteries are everywhere in your body. When a pulse is taken, it is usually taken at an artery in the neck or wrist. Explain why.

____

____

____

____

8. Identifying Relationships How are heart contractions and blood pressure related?

____

____

____

____

MATH SKILLS

9. Making Conversions An adult male's heart pumps about 2.8 million liters of blood per year. If his heart beats 70 times per minute, how much blood does his heart pump with each beat? Show your work below.

| Section Review *continued*

CHALLENGE

10. Predicting Consequences Cardiac bypass surgery allows surgeons to remove
unhealthy blood vessels near the heart. They are replaced with healthy blood
vessels from other places in the patient's body. Why might a patient need this
type of surgery?

____

____

____

____

____

____

____

____

____

____

____

____

____

____

____

____

____

Name _________________________________ Class _______________ Date ____________

Section Review

Blood

USING VOCABULARY

1. Write an original definition for *blood* and *blood pressure*.

____

____

____

____

UNDERSTANDING CONCEPTS

2. Applying A person with type B blood can donate blood to people with which type(s) of blood?

____

____

3. Describing Describe the functions of the four main components of blood.

____

____

____

4. Concluding Why is it important for a doctor to know a patient's blood type?

____

____

____

____

5. Identifying What causes blood pressure?

____

____

____

| Section Review *continued*

CRITICAL THINKING

6. Identifying Relationships How does the body use blood and blood vessels to help maintain proper body temperature?

____

____

____

____

7. Predicting Consequences Some blood diseases affect the ability of red blood cells to deliver oxygen to cells of the body. What might happen to a person with such a disease?

____

____

____

____

INTERPRETING GRAPHICS

The photomicrograph shows a WBC attacking pathogens. Use the image in your book to answer the next question.

8. Analyzing Relationships Explain the function of blood that is being demonstrated in the image.

____

____

____

MATH SKILLS

9. Making Calculations What percentage of normal (120 mm Hg) is a systolic pressure of 174 mm Hg? Show your work below.

Section Review

The Respiratory System

UNDERSTANDING CONCEPTS

1. **Describing** Describe the causes of SARS, emphysema, and some asthma attacks.

____

____

____

____

2. **Summarizing** How does breathing happen?

____

____

____

3. **Applying** Describe how your cardiovascular and respiratory systems work together.

____

____

____

____

CRITICAL THINKING

4. **Predicting Consequences** If a respiratory disorder causes lungs to fill with fluid, how could this affect a person's health?

____

____

____

____

____

| Section Review *continued*

---

5. Interpreting Statistics About 6.3 million children in the United States have asthma. About 4 million of them had an asthma attack last year. What do these statistics tell you about the relationship between asthma and asthma attacks?

____

____

____

____

____

MATH SKILLS

6. Making Calculations Usable lung capacity is about 6 L. A person can exhale about 3.6 L. What percentage of the lung capacity cannot be exhaled? Show your work below.

CHALLENGE

7. Identifying Relationships Emphysema often occurs in people who smoke cigarettes. How could smoking damage alveoli? How could emphysema affect the rest of the body?

____

____

____

____

____

____

____

Chapter Review

USING VOCABULARY

1. Academic Vocabulary In the sentence "The function of RBCs is to transport oxygen throughout the body," what does the word *function* mean?

____

Complete each of the following sentences by choosing the correct term from the word bank.

red blood cells	veins	white blood cells
arteries	pulmonary circulation	blood pressure
alveoli	bronchi	

2. ____________________________ deliver oxygen to the cells of the body.

3. ____________________________ are blood vessels that carry blood away from the heart.

4. The force that blood exerts on the walls of the arteries is

called ____________________________.

UNDERSTANDING CONCEPTS
Multiple Choice

______ **5.** _____ create blood pressure.
    a. Heart contractions   **c.** Valves
    b. Arteries   **d.** Atria

______ **6.** The parts of the heart that prevent blood from flowing backward are
    a. ventricles.   **c.** atria.
    b. capillaries.   **d.** valves.

______ **7.** Alveoli are surrounded by
    a. veins.   **c.** capillaries.
    b. muscles.   **d.** cholesterol.

______ **8.** Air moves into the lungs when the diaphragm muscle
    a. contracts and moves down.
    b. contracts and moves up.
    c. relaxes and moves down.
    d. relaxes and moves up.

| Chapter Review *continued*

Short Answer

9. Comparing Compare pulmonary circulation and systemic circulation.

____

____

10. Applying Braden's blood pressure is 110/65. What do the two numbers mean?

____

____

11. Listing List examples of one organ, one tissue, and two types of cells in the cardiovascular system.

____

____

INTERPRETING GRAPHICS

The diagram shows how the human heart would look in cross section. Use the diagram below to answer the next three questions.

12. Identifying Which letter identifies the chamber that receives blood from systemic circulation? What is this chamber's name?

____

13. Identifying Which letter identifies the chamber that receives blood from the lungs? What is this chamber's name?

____

14. Identifying Which letter identifies the chamber that pumps blood to the lungs? What is this chamber's name?

____

15. Summarizing Briefly describe the path that oxygen follows in your respiratory system and your cardiovascular system.

____

____

WRITING SKILLS

16. Writing Persuasively The blood used for blood transfusions comes from donations to blood banks. Write a paragraph persuading your friends and family members to donate blood. Use your understanding of the importance of blood transfusions in your answer.

____

____

____

____

____

____

____

____

____

____

____

____

____

____

| Chapter Review *continued*

CRITICAL THINKING

17. Concept Mapping Use the following terms to create a concept map: *blood, oxygen, alveoli, capillaries,* and *carbon dioxide.*

18. Identifying Relationships How are the functions of red blood cells and white blood cells different?

____

____

____

____

____

Chapter Review *continued*

---

19. Predicting Consequences What would happen if all of the red blood cells in your blood disappeared?

____

____

____

____

20. Applying Concepts When a person is not feeling well, a doctor may examine samples of the person's blood to see how many white blood cells are present. Why would this information be useful?

____

____

____

____

INTERPRETING GRAPHICS

Use the image below to answer the next question.

21. Making Comparisons How is the heart similar to the revolving doors in the image?

____

____

____

____

Chapter Review *continued*

22. Making Comparisons What organs are affected by heart attacks and strokes? How could this affect the organism?

____

____

____

____

____

INTERPRETING GRAPHICS

Use the image below to answer the next question.

23. Making Inferences The image shows platelets and fibers forming a blood clot to seal a broken blood vessel. A person with hemophilia cannot form blood clots normally. What do you think happens when a person with hemophilia gets a cut?

____

____

____

____

MATH SKILLS

24. Making Conversions After a person donates blood, the blood is stored in 1 pt bags until it is needed for a transfusion. A healthy person has about 5 million RBCs in each cubic millimeter (1 mm^3) of blood.

   a. How many RBCs are in 1 mL of blood? (One milliliter is equal to 1 cm^3 and to $1{,}000 \text{ mm}^3$.) Show your work below.

   b. How many RBCs are there in 1 pt? (One pint is equal to 473 mL.) Show your work below.

CHALLENGE

25. Applying Concepts Blood types are most often reported using both the ABO system and the Rh system. For example, a person with type A blood that is Rh^+ will have A^+ blood. What blood type does a person with B antigens, A antibodies, and Rh antigens have?

____

____

____

____

____

____

____

Assessment

Chapter Pretest

Teacher Notes and Answer Key

The Pretest questions are designed to help you determine the prior knowledge of your students. Some questions test whether students have mastered the background knowledge they need to understand the content you are about to teach. Other questions test your students' prior knowledge of the content you are about to teach. Use the Pretest with the Test Doctors and diagnostic teaching tips in these notes pages to help you tailor your instruction to your students' specific needs.

QUESTION NUMBER	CORRECT ANSWER	STANDARD
1	B	7.5.a
2	B	7.5.a
3	A	7.5.a
4	C	7.5.a
5	D	7.5.b
6	C	7.6.j
7	B	7.5.a
8	C	7.6.j
9	A	7.5.b
10	D	7.5.b

TEST DOCTOR

The following Pretest questions have been diagnosed by the Test Doctor. Find out what might be causing your students' "ailing" answers. Each Test Doctor is followed by a diagnostic teaching tip to help you address students' learning needs.

Question 1 *asks students to identify the type of muscle that works to move food through the digestive system.*

A Incorrect. Flexors are skeletal muscles that work to bend body parts.

B Correct. Smooth muscle tissue is found inside the organs and blood vessels.

C Incorrect. Skeletal muscles work with bones, relaxing and contracting to cause movement.

D Incorrect. Cardiac muscles are muscles found in the heart. These muscles are not under our control.

Diagnostic Teaching Tip: Students who have difficulty answering this question might benefit from creating a compare-and-contrast chart for the three types of muscles. Also include the tendons, ligament, and cartilage. Students should identify parts of the body where you would find each type of muscle. They should also indicate whether the muscle is voluntary or involuntary.

Question 2 *asks students to identify the type of exercise that is most beneficial to the heart.*

A Incorrect. Push-ups are a form of resistance exercise which is focused on strengthening skeletal muscles.

Chapter Pretest *continued*

B Correct. Jogging is a form of aerobic exercise that is focused on strengthening the heart.

C Incorrect. Lifting weights is a form of resistance exercise that is focused on strengthening skeletal muscles.

D Incorrect. Sit-ups are a form of resistance exercise that is focused on strengthening skeletal muscles.

Diagnostic Teaching Tip: Students who answer this question incorrectly should be presented with different types of exercise and asked to identify whether the exercise is focused on increasing endurance or on gaining muscle strength. This will help students identify the difference between aerobic exercise, which is good for the heart and endurance, and resistance exercise, which focuses on increasing skeletal muscle strength.

Question 3 *asks students to identify how the skeletal system helps the cardiovascular system function.*

A Correct. Red and white blood cells are made in the marrow of bones.

B Incorrect. Cardiac and smooth muscles in the blood vessels push blood through the body.

C Incorrect. Blood is not cleaned by bones but is filtered by the kidneys and the liver.

D Incorrect. Veins and arteries are not directly supported by bones. They are supported by connective tissue and the muscular system.

Diagnostic Teaching Tip: Students who have difficulty answering this question should create a listing of all the functions of bones in the skeletal system. Emphasis should be on less-obvious functions such as storage of minerals and fat, and the manufacture of blood cells. Provide a visual model to show the different types of marrow and the compact and spongy bone. Discuss where each type of bone is located in the body.

Question 4 *asks students to identify a tissue based on its function and characteristics.*

A Incorrect. Epithelial tissue covers and protects underlying tissue and is made up of epithelial cells.

B Incorrect. Nervous tissue sends electrical signals throughout the body and is made up of nerve cells.

C Correct. Muscle tissue contracts and relaxes to cause movement. It is the contraction of the heart, an organ, that causes blood pressure.

D Incorrect. Connective tissue joins, supports, insulates, nourishes, cushions, and protects organs.

Diagnostic Teaching Tip: Students who have difficulty with this question should study visuals of the different types of tissues and the cells within these tissues. Discussion should center on how the cell's structure is specialized for the particular function the cells and tissues perform.

Chapter Pretest *continued*

Question 5 *asks students to identify the body system that works with the digestive system to remove wastes from the body.*

A Incorrect. The cardiovascular system moves blood through the body but does not eliminate wastes.

B Incorrect. The endocrine system consists of hormones that control many bodily processes. It does not eliminate wastes from the body.

C Incorrect. The lymphatic system adds fluid back into the bloodstream.

D Correct. The urinary system removes unused food products and other wastes from the blood stream.

Diagnostic Teaching Tip: Students who have trouble answering this question correctly should review the body systems. For each system, they should list the major organs and the functions they perform. As an enrichment activity, students should list body systems that work with each other to help perform a vital life function. Instruction should emphasize the connections between systems and how they interact.

Question 6 *asks students to identify the function of valves in the body.*

A Incorrect. Arteries are blood vessels that carry blood away from the heart.

B Incorrect. Blood pressure is the force of blood on the inside walls of arteries.

C Correct. Valves prevent blood from going backward.

D Incorrect. Veins are blood vessels carry blood back to the heart.

Diagnostic Teaching Tip: Students who answer this question incorrectly should study a diagram of the heart and the arteries and veins connected to it. Trace the flow of blood through the heart. Point out how valves prevent blood from flowing backward. Be sure to emphasize Section 1, "The Cardiovascular System," in Chapter 16, "Circulation and Respiration." Students must know that heart valves prevent backflow of blood in the circulatory system in order to master standard 7.6.j.

Question 7 *asks students to identify what part of the blood forms a plug to help reduce blood loss at the site of an injury.*

A Incorrect. Plasma is the liquid part of blood and contains red and white blood cells as well as platelets.

B Correct. Platelets are larger than other blood cells and clump together to block the site of an injury. They then form a netting to prevent more blood loss.

C Incorrect. Red blood cells are responsible for the transport of oxygen.

D Incorrect. White blood cells destroy pathogens and other foreign materials.

Diagnostic Teaching Tip: Students who have difficulty with this question might bnenfit from knowing that blood is made up of four major components. Explain to students that when they cut themselves platelets prevent excessive blood loss. Be sure to emphasize Section 2, "Blood," in Chapter 16, "Circulation and Respiration." Students must be able to understand the function of platelets in order to master standard 7.5.a.

Chapter Pretest *continued*

Question 8 *asks students to identify how the muscles of the heart work.*

A Incorrect. No muscles work by pushing.

B Incorrect. No muscles work by twisting.

C Correct. All muscles, including those in the heart, work by contracting and relaxing.

D Incorrect. No muscles work by lengthening.

Diagnostic Teaching Tip: Students who have difficulty answering this question should discuss how the heart is mainly made of muscle. The muscle in the heart, cardiac muscle, works just like other muscles: they contract and relax in order to do their function, which is to push blood through the body. Be sure to emphasize Section 1, "The Cardiovascular System," in Chapter 16, "Circulation and Respiration." Students must know that contractions of the heart move blood through the body in order to master standard 7.6.j.

Question 9 *asks students to know how the diaphragm helps you breathe.*

A Correct. The diaphragm contracts to make the lungs inhale.

B Incorrect. The larynx guards the entrance to the trachea.

C Incorrect. The alveoli is where oxygen and carbon dioxide are exchanged.

D Incorrect. Oxygen cannot be stored until the blood is ready to take it to cells.

Diagnostic Teaching Tip: If students have difficulty with this question, they should study a diagram of the respiratory system. Discuss each of the components in the system with the class. Be sure to emphasize Section 3, "The Respiratory System," in Chapter 16, "Circulation and Respiration." Students must know that organ systems are able to function because of the contributions of individual organs in order to master standard 7.5.b.

Question 10 *asks students to determine what would happen to the cells in the human body if cellular respiration slowed down.*

A Incorrect. Coughing would not help cells with a lack of cellular respiration.

B Incorrect. Cells cannot break down carbon dioxide to get the oxygen they need.

C Incorrect. The endocrine system cannot get the oxygen the body needs.

D Correct. If cellular respiration slowed down, cells could not free energy efficiently and the person would feel tired all the time.

Diagnostic Teaching Tip: Students who have difficulty answering this question should discuss as a class why cells need oxygen (to help release the energy they need). Remind students of the role of mitochondria in cells. Be sure to emphasize Section 3, "The Respiratory System," in Chapter 16, "Circulation and Respiration." Students must know that the failure of any part of an organ system can affect the entire system in order to master standard 7.5.b.

Assessment

Chapter Pretest

______ **1.** There are different kinds of muscle tissues in your body. Some work to help you move when you run. Others do work that you are not aware of, like pumping blood through your body. What type of muscle tissue helps move the food that you eat through your digestive system?
  A flexor
  B smooth
  C skeletal
  D cardiac

______ **2.** What kind of exercise most strengthens the heart?
  A push-ups
  B jogging
  C lifting weights
  D sit-ups

______ **3.** The skeletal system does more than just support the body and protect internal organs. It also helps the cardiovascular system perform its important functions. What does the skeletal system do to help the cardiovascular system?
  A Bones make blood cells.
  B Bones help the heart push blood.
  C Blood is cleaned when it passes bones.
  D Veins and arteries are held by bone.

The chart below shows the characteristics of a tissue found in the bodies of animals and human beings.

Tissue A	
Cell Characteristics	Cells making up this tissue have proteins that are specialized for contraction.
Tissue Function	This tissue contracts and relaxes to create movement.
Organ Examples	One organ that includes Tissue A is the heart.

______ **4.** What type of tissue is Tissue A?
  A epithelial tissue
  B nervous tissue
  C muscle tissue
  D connective tissue

| Chapter Pretest *continued*

_______ **5.** Body systems work together to keep us healthy and able to survive and grow. Which of the following body systems works with your digestive system to remove unused foods from your body?
  A cardiovascular system
  B endocrine system
  C lymphatic system
  D urinary system

_______ **6.** As blood moves throughout the heart, which of the following keeps blood from flowing backwards?
  A arteries
  B blood pressure
  C valves
  D veins

_______ **7.** The blood is made up of several different components that perform different functions. What part of the blood forms a plug to help reduce blood loss when you cut your skin?
  A plasma
  B platelets
  C red blood cells
  D white blood cells

_______ **8.** How do the muscles of the heart work?
  A by pushing and pulling
  B by twisting and releasing
  C by contracting and relaxing
  D by lengthening and shortening

_______ **9.** How does the diaphragm help you breathe?
  A It contracts to make you inhale.
  B It guards the entrance to the trachea.
  C It is where oxygen and carbon dioxide are exchanged.
  D It stores oxygen until the blood is ready to take it to cells.

_______ **10.** The respiratory system takes oxygen into the body. What would happen to the cells in the human body if cellular respiration slowed down?
  A The person would begin to cough.
  B Cells would have to break down carbon dioxide to get the oxygen they need.
  C The endocrine system would have to get the oxygen the body needs.
  D Cells could not free energy efficiently, and the person would feel tired all the time.

Section Quiz

Section: The Cardiovascular System

Match the correct definition with the correct term. Write the letter in the space provided.

_______ **1.** a vessel that carries blood to the heart

_______ **2.** a collection of organs that transport blood throughout the body

_______ **3.** a tiny blood vessel that allows an exchange between blood and cells in other tissue

_______ **4.** the flow of blood from the heart to all parts of the body and back to the heart

_______ **5.** a blood vessel that carries blood away from the heart to the body's organs

_______ **6.** the flow of blood from the heart to the lungs and back to the heart

a. cardiovascular system

b. artery

c. capillary

d. vein

e. pulmonary circulation

f. systemic circulation

Write the letter of the correct answer in the space provided.

_______ **7.** What are the five main parts of the cardiovascular system?
   a. heart, arteries, capillaries, veins, and blood
   b. atherosclerosis, hypertension, heart attacks, edema, and heart failure
   c. oxygen, carbon, nitrogen, helium, and hydrogen
   d. pulmonary, systemic, lymphatic, circulatory, and respiratory

_______ **8.** What are two types of circulation of blood in the body?
   a. lymphatic and respiratory
   b. heart and blood vessels
   c. oxygen and carbon dioxide
   d. pulmonary and systemic

_______ **9.** Which cardiovascular problem results when a blood vessel in the brain becomes blocked or ruptures?
   a. stroke   **c.** heart failure
   b. heart attack   **d.** artherosclerosis

_______ **10.** Which of the following prevents blood in the cardiovascular system from going backward?
   a. atria   **c.** valves
   b. ventricles   **d.** arteries

Section Quiz

Section: Blood

Match the correct definition with the correct term. Write the letter in the space provided.

______ **1.** the force that blood exerts on the walls of arteries

______ **2.** the fluid that carries gases, nutrients, and wastes through the body

a. blood

b. blood pressure

Write the letter of the correct answer in the space provided.

______ **3.** The main components of blood are
a. systolic and diastolic.
b. antibodies, antigens, and pathogens.
c. plasma, RBCs, WBCs, and platelets.
d. A, B, AB, and O.

______ **4.** The main function of blood is to
a. digest food and break it down to be used as energy.
b. carry oxygen, nutrients, and wastes; fight disease; and regulate temperature.
c. carry messages throughout the entire body to direct body movement.
d. carry messages to parts of cells to tell them how to develop.

______ **5.** Which of the following words describe blood pressure inside large artery walls when heart ventricles contract and relax?
a. systolic and diastolic
b. antibody and antigen
c. RBC and WBC
d. plasma and platelet

______ **6.** Which of the following are blood types?
a. systolic and diastolic
b. antibodies, antigens, and pathogens
c. plasma, RBCs, WBCs, and platelets
d. A, B, AB, and O

Assessment

Section Quiz

Section: The Respiratory System

Match the correct definition with the correct term. Write the letter in the space provided.

______ **1.** the exchange of oxygen and carbon dioxide between living cells and their environment

______ **2.** a collection of organs whose primary function is to take in oxygen and expel carbon dioxide

______ **3.** the passage from the mouth to the larynx and esophagus

______ **4.** the area of the throat that contains the vocal cords and produces vocal sounds

______ **5.** the tube connecting larynx and lungs

______ **6.** two tubes connecting lungs and trachea

______ **7.** tiny air sacs of the lungs where oxygen and carbon dioxide are exchanged

a. alveoli

b. trachea

c. pharynx

d. bronchi

e. respiratory system

f. larynx

g. respiration

Write the letter of the correct answer in the space provided.

______ **8.** What are you doing when your diaphragm and rib muscles contract, and air enters the space created inside your chest cavity?
   a. coughing   **c.** inhaling
   b. eating   **d.** exhaling

______ **9.** What systems are working together during the process of respiration?
   a. respiratory and lymphatic systems
   b. respiratory and cardiovascular systems
   c. lymphatic and cardiovascular systems
   d. respiratory, lymphatic, and cardiovascular systems

______ **10.** Which of the following is a common respiratory disorder that causes bronchioles to narrow?
   a. asthma
   b. emphysema
   c. SARS
   d. hypertension

Chapter Test A

Circulation and Respiration
MULTIPLE CHOICE
Write the letter of the correct answer in the space provided.

______ **1.** What system includes the heart and blood vessels?
   a. digestive system
   b. lymphatic system
   c. cardiovascular system
   d. respiratory system

______ **2.** What does the heart do?
   a. pumps blood
   b. fights disease
   c. exchanges gases
   d. carries blood

______ **3.** What do all blood vessels do?
   a. pump blood
   b. fight disease
   c. exchange gases
   d. carry blood

______ **4.** What are the two types of circulation?
   a. systolic and diastolic
   b. pulmonary and systemic
   c. RBC and WBC
   d. plasma and platelet

______ **5.** Heart attacks and hypertension are problems of which body system?
   a. cardiovascular
   b. respiratory
   c. lymphatic
   d. digestive

______ **6.** What are alveoli?
   a. tubes that connect the throat and lungs
   b. tiny air sacs in the lungs
   c. the vocal cords in the throat
   d. the passage from the mouth to the larynx

______ **7.** What is the trachea?
   a. a tube that connects the throat and lungs
   b. the vocal cords in the throat
   c. tiny air sacs in the lungs
   d. the passage from the mouth to the larynx

Chapter Test A *continued*

_______ **8.** What is the pharynx?
   a. a tube that connects the throat and lungs
   b. the vocal cords in the throat
   c. tiny air sacs in the lungs
   d. the passage from the mouth to the larynx

_______ **9.** The process that includes inhaling, exhaling, the gain of oxygen, and the loss of carbon dioxide and water is
   a. breathing.
   b. cellular respiration.
   c. respiration.
   d. circulation.

_______ **10.** Which of the following is a respiratory disorder that causes bronchioles to narrow and restricts the body's oxygen supply?
   a. emphysema.
   b. hypertension.
   c. SARS.
   d. asthma.

MATCHING

Match the correct description with the correct term. Write the letter in the space provided.

_______ **11.** plasma, red blood cells, white blood cells, and platelets

_______ **12.** arteries, veins, and capillaries

_______ **13.** force of blood when the heart beats

_______ **14.** A, B, AB, and O

a. blood types
b. blood pressure
c. blood vessels
d. blood parts

Match the correct description with the correct term. Write the letter in the space provided.

_______ **15.** upper chamber of the heart

_______ **16.** sound made by valves

_______ **17.** lower chamber of the heart

_______ **18.** structure that stops blood from flowing backward

a. atrium
b. ventricle
c. valve
d. heartbeat

❘ Chapter Test A *continued*

FILL-IN-THE-BLANK

Use the terms from the following list to complete the sentence below.

    bronchi   larynx
    respiratory system   nose

19. Oxygen is supplied and carbon dioxide is removed by the

cardiovascular system working with the _____________________.

20. The main passageway into and out of the respiratory system

is the _____________________.

21. The area of the throat that contains the vocal cords is the

_____________________.

22. The two tubes that connect the lungs to the trachea are the

_____________________.

Chapter Test B

Circulation and Respiration
MULTIPLE CHOICE
Write the letter of the correct answer in the space provided.

______ **1.** Which of the following are the main parts of the cardiovascular system?
   a. nose, pharynx, larynx, trachea, and lungs
   b. thymus, lymph, lymph nodes, spleen, and tonsils
   c. heart, arteries, capillaries, and veins
   d. esophagus, stomach, small intestine, and large intestine

______ **2.** The function of the heart is to
   a. pump blood.
   b. fight disease.
   c. exchange gases.
   d. create energy.

______ **3.** Which of the following is the function of the three types of blood vessels?
   a. to pump blood
   b. to fight disease
   c. to exchange gases
   d. to carry blood

______ **4.** Which of the following are the two types of blood circulation in the body?
   a. systolic and diastolic
   b. pulmonary and systemic
   c. RBC and WBC
   d. plasma and platelet

______ **5.** Heart attacks, strokes, atherosclerosis, and hypertension are problems of which system?
   a. cardiovascular
   b. respiratory
   c. lymphatic
   d. digestive

______ **6.** Which of the following are the main components of blood?
   a. lymph, lymph nodes, and extracellular fluid
   b. pharynx, larynx, trachea, and bronchus
   c. plasma, red blood cells, white blood cells, and platelets
   d. A, B, AB, and O

______ **7.** Carrying oxygen and nutrients, fighting pathogens, and reducing blood
loss are some of the functions of
a. lymph.
b. blood.
c. the lungs.
d. the thymus.

______ **8.** As heart ventricles contract and then relax, blood pressure inside the
arteries changes from
a. systolic to diastolic.
b. antibody to antigen.
c. type A to type B.
d. plasma to platelet.

______ **9.** Which of the following are blood types?
a. V, W, XY, and Z
b. X, Y, XY, and Z
c. A, B, CD, and O
d. A, B, AB, and O

______ **10.** Which of the following includes breathing and the release of energy
from food?
a. systemic circulation
b. pulmonary circulation
c. circulation
d. respiration

______ **11.** Which of the following is NOT a part of the respiratory system?
a. bronchioles
b. alveoli
c. trachea
d. capillaries

______ **12.** The contraction of the diaphragm and rib muscles and expansion of
the chest cavity are part of which process?
a. cellular respiration
b. all respiration
c. inhalation
d. exhalation

______ **13.** Which of the following is NOT a respiratory disease?
a. asthma
b. emphysema
c. SARS
d. atherosclerosis

Chapter Test B *continued*

_______ **14.** A respiratory disorder that causes bronchioles to narrow and restrict
the body's oxygen supply is
a. emphysema
b. hypertension
c. SARS
d. asthma

MATCHING

**Match the correct definition with the correct term. Write the letter in the space
provided. Some terms will not be used.**

_______ **15.** a blood vessel that carries blood to the
heart

_______ **16.** a blood vessel that carries blood from the
heart

_______ **17.** air sacs in the lungs where oxygen and
carbon dioxide are exchanged

_______ **18.** the passage from the mouth to the larynx
and esophagus

_______ **19.** a group of organs that take in oxygen and
expel carbon dioxide

_______ **20.** a group of organs that transport blood
throughout the body

_______ **21.** a tiny blood vessel that allows exchanges
between blood and cells in other tissue

_______ **22.** a connective tissue that carries oxygen and
nutrients to body tissues and carries
carbon dioxide and wastes from body
tissues

a. artery

b. capillary

c. vein

d. blood

e. respiratory system

f. cardiovascular
system

g. blood pressure

h. pharynx

i. larynx

j. trachea

k. bronchus

l. alveoli

Chapter Test B *continued*

MATCHING

Match the labels to the figure of the respiratory system. Write the letters in the spaces provided.

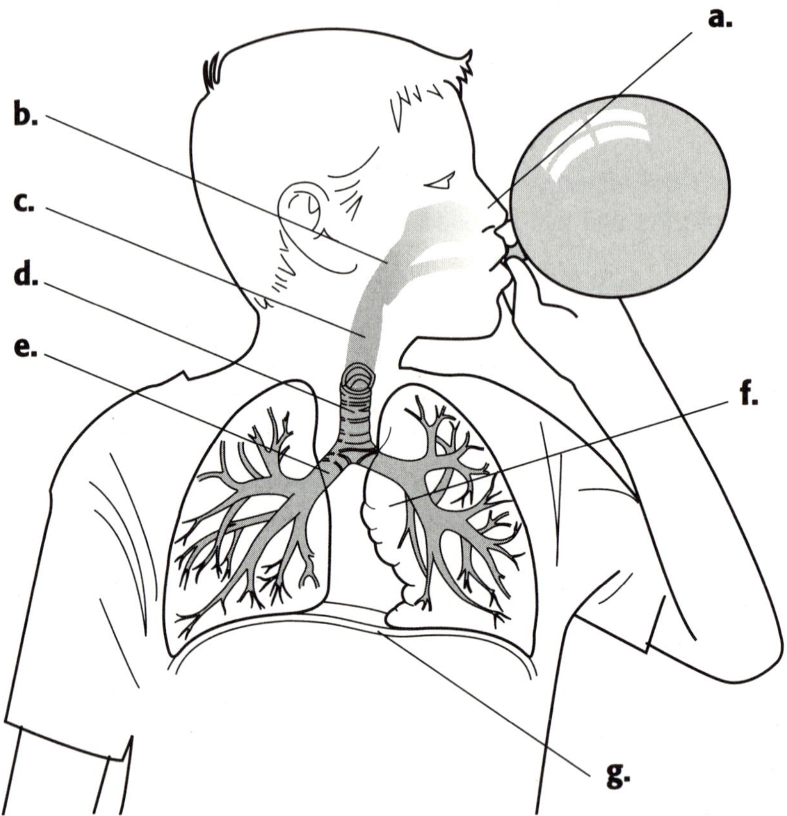

_______ **23.** larynx

_______ **24.** pharynx

_______ **25.** nose

_______ **26.** lung

_______ **27.** trachea

_______ **28.** diaphragm

_______ **29.** bronchus

Chapter Test C

Circulation and Respiration

USING KEY TERMS

Use the terms from the following list to complete the sentences below. Each term may be used only once. Some terms may not be used.

capillaries	veins	pulmonary
arteries	cellular	ventricles
systemic	blood	lungs

1. The release of energy found in molecules of glucose is known as

_______________________ respiration.

2. The lower chambers of the heart are called _______________________.

3. The movement of blood between the heart and lungs is called

_______________________ circulation.

4. Blood is carried away from the heart through

_______________________.

5. Skeletal-muscle contraction pushes blood through the

_______________________.

6. The blood vessels that enable the exchange of nutrients and oxygen between

the blood and the body's cells are _______________________.

UNDERSTANDING KEY IDEAS

Write the letter of the correct answer in the space provided.

_______ **7.** Oxygen-poor blood from the body returns to the heart's
  a. left ventricle.
  b. right ventricle.
  c. left atrium.
  d. right atrium.

_______ **8.** Which of the following is true of people with blood type AB?
  a. Their blood contains O antigens.
  b. Half of their red blood cells have A antigen; half have B antigen.
  c. Their blood can be given to a person with type O blood.
  d. They can receive blood from people with type A, type B, type AB, or
    type O blood.

_______ **9.** The exchange between alveoli and capillaries allows
    a. damaged red blood cells to be destroyed.
    b. oxygen to enter the blood.
    c. lymph to be collected.
    d. carbon dioxide to enter the blood.

_______ **10.** Red blood cells, white blood cells, and platelets are found in
    a. antibodies.
    b. plasma.
    c. hemoglobin.
    d. platelets.

_______ **11.** What do thick, elastic walls allow large arteries to do?
    a. change shape due to blood pressure
    b. store blood
    c. support large valves
    d. prevent blood from returning to the heart

_______ **12.** The pharynx is also called the
    a. larynx.
    b. windpipe.
    c. throat.
    d. lymph gland.

_______ **13.** The backflow of blood is prevented by
    a. valves.
    b. arteries.
    c. veins.
    d. atria.

14. What component of red blood cells allows these cells to carry oxygen?

____

____

15. What does the delivery of oxygen to cells allow cells to do?

____

____

16. What are the two parts of respiration?

____

____

____

17. What is the main function of the cardiovascular system?

____

____

18. Identify a cardiovascular disorder and a respiratory disorder. How do these disorders affect the body?

____

____

____

19. Describe the path that oxygen follows to enter the lungs in respiration.

____

____

____

____

CRITICAL THINKING

20. Describe what might happen if a person with type A blood receives a transfusion of type B blood.

____

____

____

21. The cardiovascular and respiratory systems are distinct. Why should they be studied together?

____

____

____

| Chapter Test C *continued*

CONCEPT MAPPING

22. Use the following terms to complete the concept map below:

antigens	platelets	blood vessels
red blood cells	plasma	pathogens

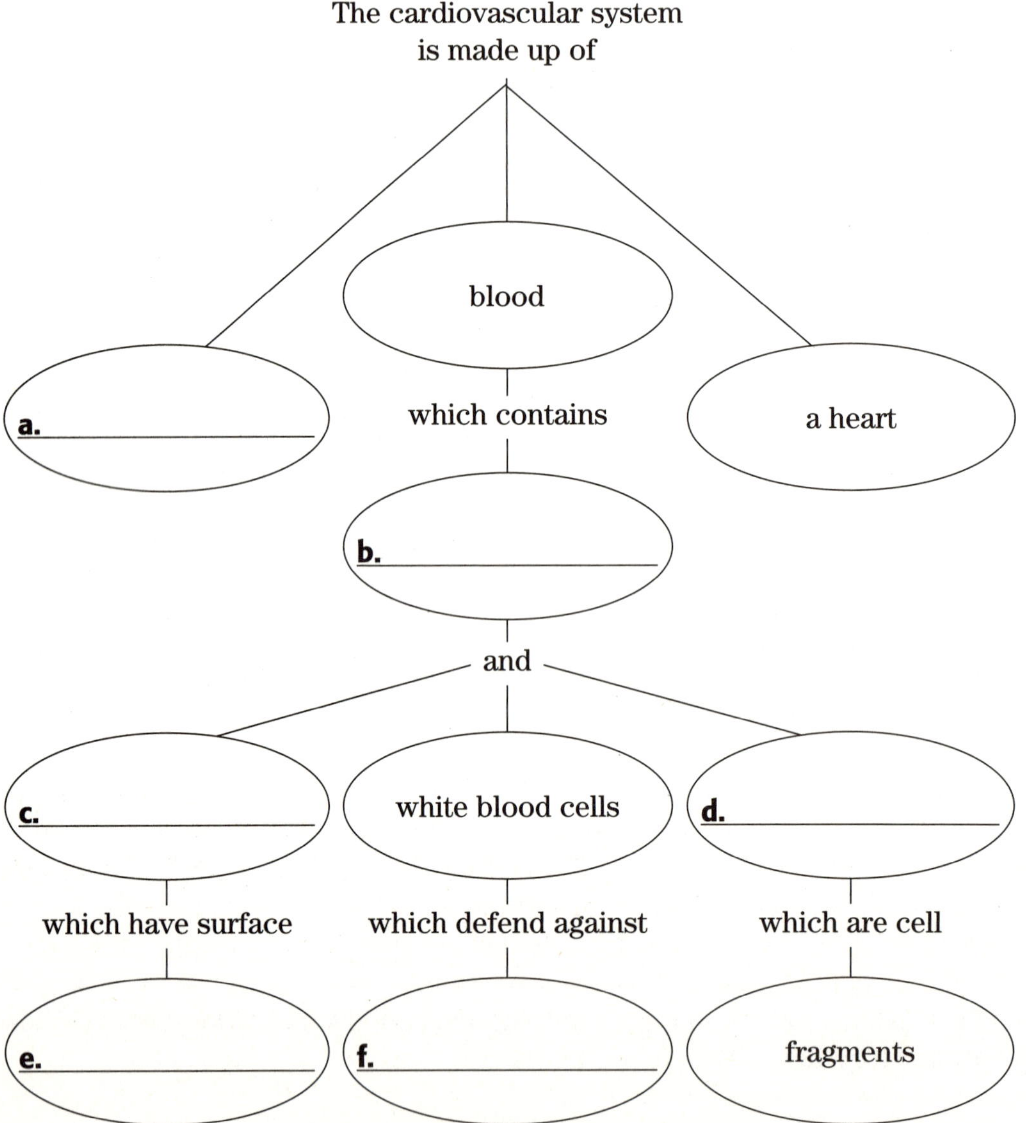

Assessment

Performance-Based Assessment

Teacher Notes

PURPOSE

Students will use simple materials to build a model of the heart and lungs. They will use their knowledge of the cardiovascular system to illustrate blood flow.

TIME REQUIRED

One 45-minute class period. Students will need 30 minutes to construct the model and 15 minutes to answer the analysis questions.

RATING

Easy ⟵ 1 2 3 4 ⟶ Hard

Teacher Prep–2
Student Set-Up–1
Concept Level–3
Clean Up–1

ADVANCE PREPARATION

Equip each student activity station with the necessary materials. Be sure to collect enough egg cartons for each group of students to have one. Also, each group will need two cardboard tubes; the tubes from bathroom tissue work best.

SAFETY INFORMATION

Remind students to use scissors safely and that pipe cleaners can have sharp ends.

TEACHING STRATEGIES

This activity works best in groups of 2–3 students. The heart will be constructed using egg-carton cups and colored pipe cleaners. The upper chambers will be labeled left and right atria, and the lower chambers will be labeled left and right ventricles. Red pipe cleaners will be used to model oxygen-rich blood flowing between the left atrium and the left ventricle and from the lungs to the left atrium. Blue pipe cleaners will be used to model oxygen-poor blood flowing between the right atrium and the right ventricle and from the right ventricle to the lungs. The cardboard tubes will be used to model the lungs.

Performance-Based Assessment *continued*

Evaluation Strategies

Use the following rubric to help evaluate student performance.

Rubric for Assessment

Possible points	Appropriate use of materials and equipment (30 points possible)
30–20	Successfully completes activity; safe handling of materials and equipment; attention to detail; superior understanding of concepts
19–10	Task is generally complete; successful use of materials and equipment; adequate observation of safety measures; sound understanding of concepts
9–1	Attempts to complete tasks yield inadequate results; sloppy use of materials and equipment; apparent lack of understanding of concepts
	Quality and accuracy of model (40 points possible)
40–30	Excellent model construction; clear and accurate labeling of structures; high level of detail
29–20	Accurate model construction; accurate labeling of structures; good level of detail
19–10	Minor inaccuracies in model construction; inaccuracies in or incomplete labeling of structures; some attention to detail
9–1	Incomplete model; inadequate or unclear labeling of structures; little attention to detail
	Explanation of observations (30 points possible)
30–20	Clear and detailed sketch of model; detailed answers to questions and superior understanding of blood flow; correct terminology in responses
19–10	Complete sketch of model; accurate answers to questions and understanding of blood flow; some correct terminology
9–1	Missing or incomplete sketch of model; inaccurate answers to questions and little understanding of blood flow; incorrect terminology used

            MODEL-MAKING

Performance-Based Assessment

OBJECTIVE

You've read about circulation and respiration in the human body. Now you will have a chance to model the blood flow in the human heart! In this activity you will use the materials provided to build a model of the heart and lungs and to illustrate blood flow.

KNOW THE SCORE!

As you work through the activity, keep in mind that you will be earning a grade for the following:

- how well you work with the materials (30%)

- how well you know the correct structure of the heart and lungs and can illustrate the blood flow between chambers (40%)

- how well you complete the analysis (30%)

MATERIALS AND EQUIPMENT

- egg carton
- glue
- pipe cleaners, red, white, and blue
- scissors
- tubes, cardboard (2)

SAFETY INFORMATION 

- Use scissors safely.
- Pipe cleaners can have sharp ends.

PROCEDURE

1. Cut the egg carton apart so that you have 8 cups. Glue pairs of them together facing each other so that you have four chambers. Label two of the chambers "atrium" and two of the chambers "ventricle." Label the atria "left" and "right." Label the ventricles "left" and "right."

2. Secure the chambers together with the pipe cleaners. Use red pipe cleaners to represent oxygen-rich blood and blue pipe cleaners to represent oxygen-poor blood. Show the blood flow into each atrium, between the atria and the ventricles, and out of each ventricle. Attach the atria and ventricles together with the white pipe cleaners.

| Performance-Based Assessment *continued*

---

3. What are blood vessels that carry blood away from the heart called?

____

____

4. What are blood vessels that carry blood toward the heart called?

____

____

5. Use the cardboard tubes to represent lungs. Connect them to the heart, again with the red and blue pipe cleaners in the correct places to represent blood vessels carrying oxygen-rich and oxygen-poor blood. Do arteries always carry oxygen-rich blood? Explain.

____

____

ANALYSIS

To answer the following questions, position your model so that it has the same orientation as the heart in your body.

6. Which side of the heart has oxygen-poor blood flowing through it?

____

7. Which side of the heart has oxygen-rich blood flowing through it?

____

8. Describe the pathway of blood going to and coming from the lungs.

____

____

9. Which of the chambers has the strongest pumping action, and why is this necessary?

____

____

BIG IDEA QUESTION

10. What are the main organ systems involved in circulation and respiration? What are some key functions that these systems work together to perform?

____

____

Assessment

Standards Assessment

Teacher Notes and Answer Key

To provide practice under more realistic testing conditions, give students 20 min to answer all of the questions in this assessment.

QUESTION NUMBER	CORRECT ANSWER	STANDARD
1	*B*	7.5 (supporting)
2	*D*	7.5.a (supporting)
3	*A*	7.6.j (supporting)
4	*B*	7.5.b (supporting)
5	*C*	7.6.j (mastering)
6	*A*	7.5.a (mastering)
7	*B*	7.5.b (mastering)
8	*D*	7.5.a (supporting)
9	*A*	7.6.b (supporting)
10	*D*	7.6.j (mastering)
11	*A*	7.5.a (supporting)
12	*C*	7.1.d (supporting)
13	*B*	7.1.d (supporting)
14	*D*	7.1.b (mastering)
15	*B*	7.2.e (mastering)

TEST DOCTOR

The following Standards Assessments questions have been diagnosed by the Test Doctor. Find out what might be causing your students' "ailing" answers. Each Test Doctor is followed by a diagnostic teaching tip to help you address students' learning needs.

Question 1 *asks students to use the correct form of the word* illustrate.

A Incorrect. The present tense of *illustrate* is not correct in this sentence.

B Correct. The complementary nature of structure and function is illustrated by the anatomy of plants and animals. The past participle of *illustrate* is needed to complete the verb phrase.

C Incorrect. The noun *illustration* does not fit in this sentence.

D Incorrect. The adjective *illustrative* does not fit this sentence.

Diagnostic Teaching Tip: Students who have difficulty with this question might benefit from reviewing parts of speech. Have students identify the parts of speech in the sentence and practice replacing each word with another word that fits the sentence grammatically.

Question 2 *asks students to match a word to its meaning.*

A Incorrect. A *process* is a set of steps, events, or changes. One example of a biological process is the process of photosynthesis.

B Incorrect. The word *system* or *organ system* refers to a group of organs working together to perform a specific function. One example of an organ system is the cardiovascular system.

C Incorrect. The *function* of a part is the use or purpose of the part. For example, the function of a valve is to stop blood from flowing backwards.

D Correct. *Structure* refers to the way something is arranged. The structure and function of an object are often related.

Diagnostic Teaching Tip: Students who have difficulty answering this question correctly might benefit from reviewing the root of the word *structure*. Challenge students to list as many words as they can think of with the root *struct* (to build) and to write a paragraph using several of them.

Question 3 *asks students to identify a synonym for the word* generate.

A Correct. The word *generate* is closest in meaning to the word *produce*. Both *generated* and *produced* are appropriate for use in the following example: Oxygen gas is ______ by the process of cellular respiration.

B Incorrect. *Find* means "to discover after a search."

C Incorrect. *Filter* means "to separate particles from a gas or liquid."

D Incorrect. *Destroy* means "to demolish or ruin." *Destroy* and *generate* are antonyms.

Diagnostic Teaching Tip: Students who have difficulty with this question might benefit from reviewing words similar to *generate*. Challenge students to think of five words with similar meanings, and then, ask them to use each word in a sentence that illustrates how it has a slightly different meaning from the others.

Question 4 *asks students to match a word to its definition.*

A Incorrect. A *distribution* is the scattering of something over an area.

B Correct. A *contribution* is something that is given. Body systems work through the contributions of cells, tissues, and organs.

C Incorrect. *Execution* is the carrying out of an action.

D Incorrect. *Retribution* means "something given or demanded as repayment."

Diagnostic Teaching Tip: Students who struggle to answer this type of question might benefit from reviewing the meaning of *contribution*. Instruct students to think of different contexts in which the word is used and then, to define their common ground.

Standards Assessment *continued*

Question 5 *asks students to demonstrate understanding of blood flow in the cardiovascular system.*

A Incorrect. Expansion of the heart muscle, which occurs when the heart relaxes, does not cause blood to pump.
B Incorrect. Blood vessel expansion, commonly called blood vessel dilation, occurs within blood vessels close to the skin when the body is hot. Heat is released to the environment.
C Correct. When the heart muscle contracts, blood is pumped into the cardiovascular system.
D Incorrect. Blood does not flow as a result of blood vessels contracting.

Diagnostic Teaching Tip: Students who have difficulty answering this question correctly might benefit from reviewing the steps of circulation. Give students a blank diagram of the human circulatory system. Have them draw arrows on the diagram to show the direction of blood flow through the heart and the path blood takes through the body.

Question 6 *asks students to demonstrate understanding of the components of the circulatory system.*

A Correct. Capillaries are the smallest blood vessels. They are rarely more than three cells away from any cell. The thin walls of capillaries allow for the exchange of gases, nutrients, and wastes.
B Incorrect. Arteries carry blood away from the heart to capillaries.
C Incorrect. Ventricles are parts of the heart, not blood vessels.
D Incorrect. Veins are blood vessels that carry blood back to the heart from capillaries.

Diagnostic Teaching Tip: Students who have difficulty answering this question correctly might benefit from reviewing the different types of blood vessels and their functions. Work with students to create a three-column chart that identifies and describes the role and characteristics of veins, arteries, and capillaries in the body.

Question 7 *asks students to demonstrate understanding of how body systems function.*

A Incorrect. This is an example of atherosclerosis.
B Correct. When a blood vessel ruptures in the brain, the cells surrounding or past the damaged vessel are unable to get blood. The resulting lack of oxygen can lead to a stroke.
C Incorrect. Heart failure occurs when the heart cannot pump enough blood for the body's needs.
D Incorrect. A heart attack occurs when oxygen is cut off to the heart.

Diagnostic Teaching Tip: Students who have difficulty with this question might benefit from reviewing the causes of problems related to the cardiovascular system. Describe a set of cardiovascular symptoms, and have students try to identify the affliction being described.

Question 8 *asks students to demonstrate knowledge of the parts of the respiratory system.*

A Incorrect. Capillaries are part of the cardiovascular system.

B Incorrect. Though the larynx is part of the respiratory system, it is not located within the area shown in the diagram.

C Incorrect. Though the pharynx is part of the respiratory system, it is not located within the area shown in the diagram.

D Correct. The bronchi branch into smaller passageways, called *bronchioles*, in the lungs.

Diagnostic Teaching Tip: Students who have difficulty answering this question correctly might benefit from reviewing the parts of the respiratory system. Have students draw their own diagrams of the respiratory system, making certain to label the bronchus, bronchioles, larynx, pharynx, trachea, lungs, alveoli, nose, and diaphragm.

Question 9 *asks students to identify how the cardiovascular system regulates body temperature.*

A Correct. When body temperature rises, the brain tells the blood vessels in the skin to enlarge, releasing heat through the skin into the environment.

B Incorrect. When blood vessels in the skin contract, the action prevents the release of heat through the skin to the environment. Blood vessels contract when the body becomes too cold.

C Incorrect. A temperature change does not lead to a decrease in heart rate.

D Incorrect. A temperature change does not lead to an increase in heart rate. An increase in heart rate may be caused by many other stimuli, such as exercise or stress.

Diagnostic Teaching Tip: Students who have difficulty with this question might benefit from reviewing what factors are regulated by the cardiovascular system. Invite students to study an anatomy chart of the cardiovascular system. Have them describe how the system works, based on what they see in the chart. Then, discuss the functions that are not very obvious, such as regulation of bleeding, body temperature, and blood pressure.

Standards Assessment *continued*

Question 10 *asks students to demonstrate understanding of the types of blood pressure.*

A Incorrect. Blood pressure is not measured in capillaries.
B Incorrect. The pressure inside arteries when the heart relaxes is diastolic blood pressure.
C Incorrect. Blood pressure is not measured in veins.
D Correct. Systolic pressure is the pressure on arterial walls when the heart contracts. Blood pressure is often reported as a ratio of systolic to diastolic blood pressure, and in units of mm Hg.

Diagnostic Teaching Tip: Students who have difficulty answering this question correctly might benefit from creating pamphlets explaining blood pressure and ways to maintain healthy blood pressure. Students should present their pamphlets to other students.

Question 11 *asks students to demonstrate knowledge of the cardiovascular system.*

A Correct. Part A in the diagram is an atrium.
B Incorrect. Part B is a ventricle, not an atrium.
C Incorrect. Part C is a valve.
D Incorrect. Part D is an artery leading out of the heart.

Diagnostic Teaching Tip: Students who have difficulty answering this question correctly might benefit from creating an unlabeled diagram of the heart. Instruct students to write labels and descriptions of each part of the heart on separate sticky notes. Students can challenge each other to match the correct sticky notes to the parts of the heart.

Question 12 *asks students to determine the starting materials of cell processes.*

A Incorrect. Fermentation is a process that releases the energy stored in food without using oxygen.
B Incorrect. Cell division is a process that results in two identical daughter cells.
C Correct. Cellular respiration is a process that releases the energy stored in food using oxygen. Cellular respiration is much more efficient at releasing energy than fermentation.
D Incorrect. Meiosis is the process that results in haploid sex cells.

Diagnostic Teaching Tip: Students who have difficulty answering this question correctly might benefit from reviewing cellular processes. Divide students into groups, and have each group make a poster that shows a cellular process. Hang the posters around the classroom.

Standards Assessment *continued*

Question 13 *asks students to describe cellular respiration.*

A Incorrect. Sunlight and carbon dioxide are converted into food, which contains energy, during photosynthesis.

B Correct. The process of cellular respiration uses oxygen and sugar molecules to release the energy stored in the food.

C Incorrect. Sunlight is used to make sugar molecules by the process of photosynthesis. Oxygen is released as a waste product.

D Incorrect. Sunlight is not part of the process of cellular respiration.

Diagnostic Teaching Tip: Students who have trouble with this question might be confusing the processes, reactants, and products of cellular respiration and photosynthesis. Write down the chemical equations for the two processes, pointing out that cellular respiration provides the materials needed for photosynthesis to begin: carbon dioxide and water. The process of photosynthesis, in turn, provides the materials needed for cellular respiration: sugar and oxygen. Have students create a diagram that illustrates how the two processes form a cycle.

Question 14 *asks students to demonstrate understanding of the differences between plant and animal cells.*

A Incorrect. A blood cell, like all animal cells, has mitochondria.

B Incorrect. Every cell has a cell membrane.

C Incorrect. A white blood cell has a nucleus, though mature red blood cells do not.

D Correct. Blood cells, like other animal cells, do not have chloroplasts.

Diagnostic Teaching Tip: Students who have difficulty with this question might benefit from reviewing the parts of animal and plant cells. Have students work with a partner to identify the parts on unlabeled diagrams or models of a plant and an animal cell.

Question 15 *asks students to demonstrate knowledge that DNA contains the genetic material of a cell.*

A Incorrect. Ribosomes are the site of protein synthesis. They do not contain genetic material.

B Correct. DNA is the genetic material for every cell.

C Incorrect. The endoplasmic reticulum makes, packages, and moves lipids and proteins. Genetic material is not found in endoplasmic reticulum.

D Incorrect. A nucleolus may be found in the nuclei of some cells. The nucleolus is the site of the first part of ribosome synthesis.

Diagnostic Teaching Tip: Students who have difficulty answering this question correctly might benefit from reviewing the functions of different cell parts. Have students list the parts of a cell on the left side of a chart, and then, fill in the right side with a description of its function.

Standards Assessment

REVIEWING ACADEMIC VOCABULARY

______ **1.** Choose the appropriate form of the word *illustrate* for the following
sentence: The complementary nature of structure and function is
______ by the anatomy of plants and animals.
A illustrate
B illustrated
C illustration
D illustrating

______ **2.** Which of the following words means "the arrangement of the parts of
the whole"?
A process
B system
C function
D structure

______ **3.** Which of the following words is the closest in meaning to the word
generate?
A produce
B find
C filter
D destroy

______ **4.** Which of the following words means "a part given to the whole"?
A distribution
B contribution
C execution
D retribution

REVIEWING CONCEPTS

______ **5.** Which of the following actions pumps blood through the body?
A the expansion of the heart
B the expansion of blood vessels
C the contractions of the heart
D the contractions of the blood vessels

______ **6.** Which type of blood vessel allows for the exchange of gases between
the blood and body cells?
A capillaries
B arteries
C ventricles
D veins

Standards Assessment *continued*

_______ **7.** Which of the following is a direct cause of a stroke?
 A Cholesterol builds up in an artery and blocks blood flow through it.
 B A blood vessel ruptures in the brain, so oxygen flow is cut off.
 C The heart cannot pump enough blood to meet the body's needs.
 D A blood vessel collapses, so the flow of oxygen to the heart is cut off.

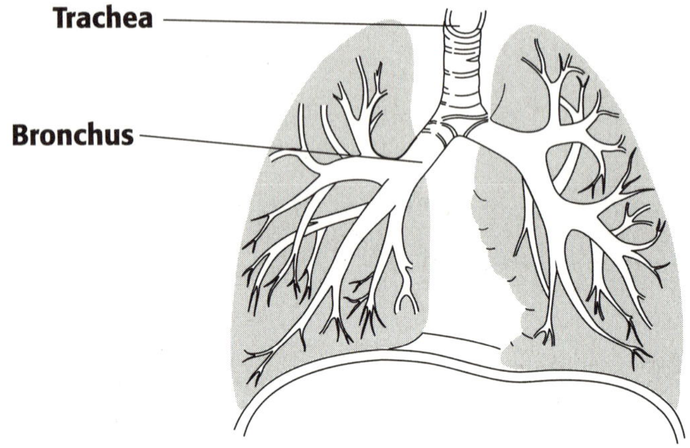

Trachea

Bronchus

_______ **8.** The diagram above shows parts of the respiratory system. What label could be added to the diagram?
 A Capillaries
 B Larynx
 C Pharynx
 D Bronchioles

_______ **9.** Which of the following occurs when the body becomes too warm and tries to cool itself?
 A Blood vessels in the skin enlarge.
 B Blood vessels in the skin contract.
 C The heart pumps blood more slowly.
 D The heart pumps blood more quickly.

_______ **10.** What is systolic blood pressure?
 A The pressure inside capillaries when the heart contracts.
 B The pressure inside arteries when the heart relaxes.
 C The pressure inside veins when the heart contracts.
 D The pressure inside arteries when the heart contracts.

Standards Assessment *continued*

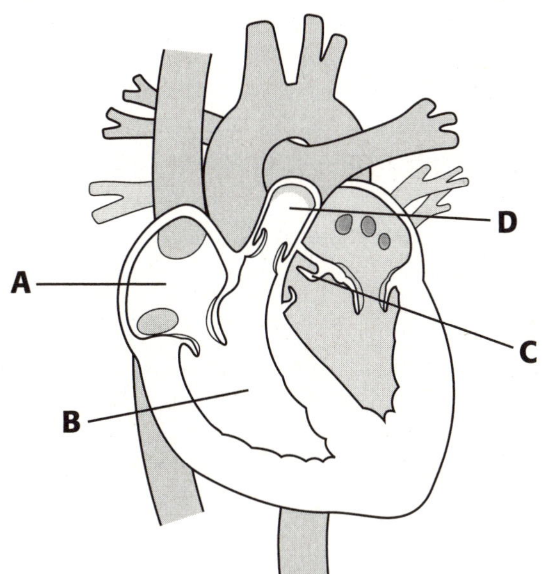

_______ **11.** Which part of the diagram above shows an atrium?
A part A
B part B
C part C
D part D

REVIEWING PRIOR LEARNING

_______ **12.** Why do cells need oxygen?
A Oxygen is used during the process of fermentation.
B Oxygen is used during the process of cell division.
C Oxygen is used during the process of cellular respiration.
D Oxygen is used during the process of meiosis.

_______ **13.** What is the result of cellular respiration?
A Energy is produced from radiant sunlight and carbon dioxide.
B Energy is produced from sugar molecules and oxygen.
C Sunlight is converted into sugar molecules and oxygen.
D Sunlight is converted into water molecules.

_______ **14.** A white blood cell is not likely to have a
A mitochondrion.
B cell membrane.
C nucleus.
D chloroplast.

_______ **15.** Which of the following is the genetic material for a cell?
A ribosome
B DNA
C endoplasmic reticulum
D nucleolus

  DATASHEET

Modeling a Valve

Teacher Notes

In this activity, students will create a model of a one-way valve (covers standards 7.6.j and 7.7.d). Cardboard tubes from paper towel rolls are ideal to use as models of blood vessels. Otherwise, you can make tubes by rolling up sheets of paper. Suggest that students add the marbles slowly. Otherwise, a rush of marbles might tear or dislodge the paper cone.

MATERIALS

For each group

- bowl
- marbles (handful)
- paper (1 sheet)
- scissors
- tape
- tube, cardboard, from paper towel roll OR a second sheet of paper and tape

SAFETY CAUTION

Remind students to review all safety cautions and icons before beginning this activity.

Modeling a Valve

In this activity, you will create a model of a valve that allows material to flow in only one direction. Similar valves in the heart and in certain large blood vessels prevent blood from flowing in the wrong direction.

SAFETY INFORMATION 

PROCEDURE

1. Roll a **sheet of paper** into a cone
   - Secure it with a piece of **tape.**
   - The diameter of the wide part of the cone should be 7 cm.
   - Make a series of paper flaps: Use **scissors** to create several slits that extend outward from the point of the cone,.

2. Insert the point of the cone into one end of a **cardboard tube.**
   - Secure the tube in place with tape.
   - Place a **bowl** beneath the open end of the tube.

3. Hold the tube with the cone positioned at the top of the tube.
   - Slowly pour a handful of **marbles** into the tube.
   - Did the marbles go through?

____

4. Turn the tube upside down, so that the cone is at the bottom.
   - Pour the marbles into the tube.
   - Did the marbles go through?

____

ANALYSIS

5. When you poured the marbles through the tube, could the marbles move through the tube in both directions?

____

____

____

Modeling a Valve *continued*

6. When the marbles encounter the wide end of the cone first, what happens to the flaps? Can the marbles flow through the passage?

____

____

When the marbles encounter the narrow end of the cone first, what happens to the flaps? Can the marbles flow through the passage?

____

____

7. If the heart did not have valves, what could happen to the flow of blood in the heart?

____

____

Explore Activity) **DATASHEET B**

Modeling a Valve

In this activity, you will create a model of a valve that allows material to flow in only one direction. Similar valves in the heart and in certain large blood vessels prevent blood from flowing in the wrong direction.

SAFETY INFORMATION 

PROCEDURE

1. Roll a **sheet of paper** into a cone, and secure it with a piece of **tape.** The diameter of the wide part of the cone should be 7 cm. Use **scissors** to create several slits that extend outward from the point of the cone so that you make a series of paper flaps.

2. Insert the point of the cone into one end of a **cardboard tube.** Secure the tube in place with tape. Place a **bowl** beneath the open end of the tube.

3. Hold the tube with the cone positioned at the top. Pour a **handful of marbles** into the tube.

4. Turn the tube upside down so that the cone is at the bottom. Pour the marbles into the tube.

ANALYSIS

5. How did the paper cone affect the movement of the marbles?

____

____

6. Describe how this model valve works.

____

____

7. Why do you think valves are needed in the heart?

____

____

Explore Activity

DATASHEET C

Modeling a Valve

In this activity, you will create a model of a valve that allows material to flow in only one direction. Similar valves in the heart and in certain large blood vessels prevent blood from flowing in the wrong direction.

SAFETY INFORMATION

PROCEDURE

1. Roll a **sheet of paper** into a cone and secure it with a piece of **tape.** The diameter of the wide part of the cone should be 7 cm. Use **scissors** to create several slits that extend outward from the point of the cone, so that you make a series of paper flaps.

2. Insert the point of the cone into one end of a **cardboard tube.** Secure the tube in place with tape. Place a **bowl** beneath the open end of the tube.

3. Hold the tube with the cone positioned at the top. Pour a handful of **marbles** into the tube.

4. Turn the tube upside down so that the cone is at the bottom. Pour the marbles into the tube.

ANALYSIS

5. Describe the effect of the paper cone on the movement of marbles through the tube.

____

____

6. Describe how this model valve works.

____

____

7. Explain the function of valves in the heart and large blood vessels.

____

____

Quick Lab

DATASHEET

Vessel Blockage

Teacher Notes

This activity will help students understand how a blockage in a blood vessel can affect the whole organism (covers standard 7.5.b).

MATERIALS

For each group

- facial tissue
- straws (4 or 5)
- tape, clear
- wire, small piece

SAFETY CAUTION

Remind students to review all safety cautions and icons before beginning this activity.

Quick Lab   **DATASHEET A**

Vessel Blockage

SAFETY INFORMATION

PROCEDURE

1. Connect **4 or 5 straws** together by taping them end to end with **clear tape.**

2. Tape a **tissue** to the side of your desk so that it hangs.
   - Blow through your long straw so that the tissue moves.

3. Tightly wrap a **small piece of wire** around one section of your straw.
   - Blow through your straw and try to make the tissue move.
   - Does the tissue move? How does the wire blockage affect the function of the straw?

   ____

   ____

4. In atherosclerosis, blood vessels become narrow and less elastic due to buildup of cholesterol and other lipids. Blood does not flow easily through these vessels.
   - How is wrapping wire around the straw similar to atherosclerosis?

   ____

   ____

   ____

5. All cells in the body rely on blood and blood vessels to supply nutrients and carry away wastes.
   - How might the failure of one blood vessel in the body affect the rest of the organism?

   ____

   ____

   ____

DATASHEET B

Vessel Blockage

SAFETY INFORMATION

PROCEDURE

1. Connect **4 or 5 straws** together by taping them end to end with **clear tape.**

2. Tape a **tissue** to the side of your desk so that the tissue hangs. Blow through your long straw so that the tissue moves.

3. Tightly wrap a **small piece of wire** around one section of your straw. Blow through your straw, and try to make the tissue move. How does this blockage affect the function of the straw?

____

____

____

4. How does this activity relate to atherosclerosis?

____

____

____

5. How might the failure of one blood vessel in the body affect the rest of the organism?

____

____

____

____

Name _______________________________ Class _______________ Date ____________

                                    DATASHEET C

Vessel Blockage

SAFETY INFORMATION

PROCEDURE

1. Connect **4 or 5 straws** together by taping them end to end with **clear tape.**

2. Tape a **tissue** to the side of your desk so that it hangs. Blow through your long straw so that the tissue moves.

3. Tightly wrap a **small piece of wire** around one section of your straw. Blow through your straw and try to make the tissue move. Describe the effect of the blockage on the function of the straw.

   ____

   ____

4. Explain how this experiment models atherosclerosis.

   ____

   ____

5. Describe the effects of the failure of one blood vessel in the body on the rest of an organism.

   ____

   ____

   ____

  DATASHEET

Modeling Blood Pressure

Teacher Notes

In this activity, students will demonstrate systolic and diastolic blood pressure in an artificial apparatus (covers standards 7.6.j and 7.7.d). You can use laboratory film, such as the film used to seal petri dishes, in place of tape to attach the balloon to the pipet bulb.

MATERIALS

For each group

- balloon, long
- pipet bulb
- tape
- water

SAFETY CAUTION

Caution students to pay close attention while attaching the balloon to the pipet bulb, so that the balloon does not snap away and hit someone.

Modeling Blood Pressure

In this activity, you will demonstrate systolic and diastolic blood pressure. You will use a pipet bulb to represent the heart.

PROCEDURE

1. Fill a **pipet bulb** with water.
   - Stretch the mouth of a long balloon around the end of the pipet bulb.
   - Secure with tape.

2. Carefully squeeze the pipet bulb in one hand.
   - Look at the balloon. Does the pressure in the balloon increase?

____

3. Release your squeeze on the pipet bulb.
   - Look at the balloon. Does the pressure in the balloon decrease?

____

4. If the pipet bulb represents the heart, what does the balloon represent?

____

5. Systolic pressure is the pressure in the arteries when the heart ventricles contract. Which state, bulb squeezed or not squeezed, is similar to systolic pressure? Explain.

____

____

6. If your systolic pressure is 95 mm Hg and your diastolic pressure is 60 mm

Hg, your blood pressure is _____________________ .

Modeling Blood Pressure

In this activity, you will demonstrate systolic and diastolic blood pressure. You will use a pipet bulb to represent the heart.

PROCEDURE

1. Fill a **pipet bulb** with **water.** Stretch the mouth of a **long balloon** around the end of the pipet bulb. Secure with **tape.**

2. Carefully squeeze the pipet bulb in one hand. Describe the pressure in the balloon.

   ____

3. Release your squeeze on the pipet bulb. Describe the pressure in the balloon now.

   ____

4. If the pipet bulb represents the heart, what does the balloon represent?

   ____

   ____

5. Which state, bulb squeezed or not squeezed, is similar to systolic pressure? Explain.

   ____

   ____

   ____

6. What is your blood pressure if your diastolic pressure is 60 mm Hg and your systolic pressure is 95 mm Hg?

   ____

   ____

   ____

Quick Lab

Modeling Blood Pressure

In this activity, you will demonstrate systolic and diastolic blood pressure. You will use a pipet bulb to represent the heart.

PROCEDURE

1. Fill a **pipet bulb** with water. Stretch the mouth of a **long balloon** around the end of the pipet bulb. Secure with **tape.**

2. Carefully squeeze the pipet bulb in one hand. Describe the pressure in the balloon.

   ____

3. Release your squeeze on the pipet bulb. Describe the pressure in the balloon now.

   ____

4. Your setup models the heart and arteries. What does the pipet bulb represent? What does the balloon represent?

   ____

   ____

5. Which state of the bulb in the procedure is similar to systolic pressure? Explain.

   ____

   ____

6. What is your blood pressure if your diastolic pressure is 60 mm Hg and your systolic pressure is 95 mm Hg?

   ____

(Quick Lab)

Replicate Respiration

Teacher Notes

This activity will help students understand the complex interactions of organs that cause inhalation (covers standard 7.5.a).

MATERIALS

For each group

- balloon
- glove, latex
- scissors
- soda bottle, 1 L, empty
- tape

SAFETY CAUTION

Remind students to review all safety cautions and icons before beginning this activity.

Quick Lab

DATASHEET A

Replicate Respiration

SAFETY INFORMATION

PROCEDURE

1. Use **scissors** to cut the bottom off an empty **1 liter soda bottle.**
   - Discard the bottom.

2. Stretch a **latex glove** over the cut end of the bottle.
   - Secure the glove with tape.

3. Insert a **balloon** into the top of the bottle.
   - Pull the lip of the balloon to fit over the mouth of the bottle.
   - Secure the balloon with tape.

4. Pull on the fingers of the glove.
   - Look at the balloon. Does the balloon expand?

   ____

5. During respiration, the diaphragm moves down and creates a vacuum in the chest that sucks air into the lungs.
   - How is the glove like the diaphragm during respiration?

   ____

   - How is the balloon like the lungs during respiration?

   ____

   ____

   ____

Quick Lab

Replicate Respiration

SAFETY INFORMATION

PROCEDURE

1. Use **scissors** to cut the bottom off an empty **1 liter soda bottle.**

2. Stretch a **latex glove** over the open bottom of the bottle. Secure with **tape.**

3. Insert a **balloon** into the top of the bottle. Pull the lip of the balloon to fit over the mouth of the bottle. Secure with tape.

4. Pull on the fingers of the glove. What happens to the balloon?

____

____

____

____

____

5. How does this activity relate to respiration and the diaphragm?

____

____

____

____

____

Name _______________________________ Class _______________ Date ___________

Replicate Respiration

SAFETY INFORMATION

PROCEDURE

1. Use **scissors** to cut the bottom off an empty **1 liter soda bottle.**

2. Stretch a **latex glove** over the open bottom of the bottle. Secure with **tape.**

3. Insert a **balloon** into the top of the bottle. Pull the lip of the balloon to fit over the mouth of the bottle. Secure with tape.

4. Pull on the fingers of the glove. Describe and explain the effects on the balloon.

____

____

5. Explain how this activity models respiration.

____

____

____

____

____

Skills Practice Lab)   **DATASHEET**

Carbon Dioxide in Respiration

Teacher Notes

In this lab, students will detect the presence of carbon dioxide in their breath, record the resulting data, and compare this data with that of their classmates (covers standard 7.7.c). Students will then describe how various parts of the body share the task of removing gaseous wastes (covers standards 7.5.a and 7.5.b).

Sushma Kashyap, MSc
Alvarado
Intermediate School
Rowland Heights, CA

TIME REQUIRED

One 45-minute class period

LAB RATINGS

Easy ←——— 1   2   3   4 ——→ Hard

Teacher Prep–2
Student Set-Up–2
Concept Level–2
Clean Up–2

MATERIALS

The materials listed on the student page are enough for two students. You may wish to substitute bromothymol blue indicator solution for the phenol red indicator solution. The bromothymol blue solution will turn green in the presence of carbon dioxide. Use eight drops of bromothymol blue indicator to make the solution. Clear plastic cups (6 oz or 8 oz) may be used instead of 250 mL flasks if glassware is in short supply or if you have concerns about breakage.

SAFETY CAUTION

Remind students to review all safety cautions and icons before beginning this lab activity. Tell students to inform you if any glassware is broken. Remind students that they should only exhale through the straw.

PREPARATION NOTES

Tell students that carbon dioxide is in the air in the classroom. Tell them to make sure that they do not expose their indicator solution to air for several minutes before they use it in the experiment.

Carbon Dioxide in Respiration

Carbon dioxide is important to both plants and animals. Plants take in carbon dioxide during photosynthesis and give off oxygen as a byproduct of the process. Animals—including you—take in oxygen during cellular respiration and give off carbon dioxide as a byproduct of the process.

OBJECTIVES

Detect the presence of carbon dioxide in your breath before and after exercise.

Compare the data for carbon dioxide in your breath with the data from your classmates.

MATERIALS

- eyedropper
- flask, Erlenmeyer, 250 mL (2)
- gloves, protective
- graduated cylinder, 100 mL
- paper towels
- phenol red indicator solution
- stopwatch, or clock with a second hand
- straw, plastic drinking (2)
- water, 200 mL

SAFETY INFORMATION 

PROCEDURE

1. Find a partner.
   - Put on your gloves, safety goggles, and apron.

2. Use the graduated cylinder to pour 100 mL of water into a 250 mL flask. (Hint: Make sure to look at eye level when you read the volume.)

3. Repeat step 2 so that you have two flasks, each containing 100 mL of water.

4. Using an eyedropper, carefully place four drops of phenol red indicator solution into the water of the first flask.
   - The water should turn orange.

5. Place a plastic drinking straw into the solution of phenol red and water.
   - Drape a paper towel over the flask to prevent splashing.

6. Make sure your lab partner has the stopwatch ready.
   - You will soon exhale into the straw.
   - Your lab partner should begin keeping time as soon as you start to exhale through the straw.

Carbon Dioxide in Respiration *continued*

7. Breathe normally, but exhale into the straw in the solution.
   - **Caution:** Do not inhale through the straw. Do not drink the solution, and do not share a straw with anyone.
   - Have your lab partner time how long the solution takes to change color.
   - On the line below, record the time the solution takes to change color.

____

8. Do jumping jacks or sit-ups for 3 min.
   - Repeat steps 4–7 using the second flask.
   - Record the time the solution takes to change color.

____

ANALYZE THE RESULTS

9. Describing Events The phenol red started out orange.
   - Did the indicator solution change color?

____

   - What color did it change to?

____

10. Organizing Data Use the table to record your data.
   - Write your data in the appropriate columns.
   - Collect data from your classmates and add this to your table.

Student name	Time without exercise	Time with exercise

Carbon Dioxide in Respiration *continued*

11. Examining Data Look at the data in the first column, "Time without exercise."
- What was the longest length of time it took to see a color change?

____

- What was the shortest?

____

- How do you account for the difference?

____

____

- Look at the data in the second column, "Time with exercise."
- What was the longest length of time it took to see a color change?

____

- What was the shortest?

____

- How do you account for the difference?

____

12. Examining Data Look at your previous answers.
- Did exercising make a difference in the time it took for the solution to change color?

____

- Why or why not?

____

____

13. Constructing Graphs In the space below, make a bar graph that compares your data from the first column with the data of your classmates.
- Put time to see color change on the y-axis and student names on the x-axis.

Carbon Dioxide in Respiration *continued*

DRAW CONCLUSIONS

14. Interpreting Information Do you think the solution would change color more quickly for males or females? Explain your answer.

____

____

- Do you think the solution would change color more quickly for athletic students? Explain your answer.

____

____

BIG IDEA QUESTION

15. Making Predictions Which individual organs in the body play a role in getting rid of waste products, such as carbon dioxide?

____

- If these organs failed and excess carbon dioxide could not be removed from the body, what might happen to the respiratory system and to the person?

____

____

Carbon Dioxide in Respiration

Carbon dioxide is important to both plants and animals. Plants take in carbon dioxide during photosynthesis and give off oxygen as a byproduct of the process. Animals—including you—take in oxygen during cellular respiration and give off carbon dioxide as a byproduct of the process.

OBJECTIVES

Detect the presence of carbon dioxide in your breath before and after exercise.

Compare the data for carbon dioxide in your breath with the data from your classmates.

MATERIALS

- eyedropper
- flask, Erlenmeyer, 250 mL (2)
- gloves, protective
- graduated cylinder, 100 mL
- paper towels
- phenol red indicator solution
- stopwatch, or clock with a second hand
- straw, plastic drinking (2)
- water, 200 mL

SAFETY INFORMATION 

PROCEDURE

1. Find a partner. Put on your gloves, safety goggles, and apron.

2. Use the graduated cylinder to pour 100 mL of water into a 250 mL flask.

3. Repeat step 2 so that you have two flasks, each containing 100 mL of water.

4. Using an eyedropper, carefully place four drops of phenol red indicator solution into the water of the first flask. The water should turn orange.

5. Place a plastic drinking straw into the solution of phenol red and water. Drape a paper towel over the flask to prevent splashing.

6. Make sure your lab partner has the stopwatch ready. You will soon exhale into the straw. Your lab partner should begin keeping time as soon as you start to exhale through the straw.

7. Breathe normally, but exhale into the straw in the solution. **Caution:** Do not inhale through the straw. Do not drink the solution, and do not share a straw with anyone. Have your lab partner time how long the solution takes to change color. Record the time.

8. Do jumping jacks or sit-ups for 3 min, and repeat steps 4–7 using the second flask.

Carbon Dioxide in Respiration *continued*

ANALYZE THE RESULTS

9. Describing Events Describe what happened to the indicator solution.

____

____

10. Organizing Data Make a table with three columns. Title the columns "Student name," "Time without exercise," and "Time with exercise." Write your data in the appropriate columns. Collect data from your classmates, and add the data to your table.

11. Examining Data Look at the data in the first column. What was the longest length of time it took to see a color change? What was the shortest? How do you account for the difference? Repeat for the second column.

____

____

____

____

____

12. Examining Data Did exercising make a difference in the time it took for the solution to change color? Why or why not?

____

____

____

____

____

13. Constructing Graphs Make a bar graph that compares your data from the first column with the data of your classmates.

____

____

____

____

____

____

DRAW CONCLUSIONS

14. Interpreting Information Do you think that there is a relationship between the length of time the solution takes to change color and the person's physical characteristics, such as which gender the tester is or whether the tester has an athletic build? Explain.

____

____

____

____

____

____

BIG IDEA QUESTION

15. Making Predictions Which individual organs in the body contribute to getting rid of waste products, such as carbon dioxide? What would happen if these organs failed and excess carbon dioxide could not be removed from the body?

____

____

____

____

____

Skills Practice Lab   **DATASHEET C**

Carbon Dioxide in Respiration

Carbon dioxide is important to both plants and animals. Plants take in carbon dioxide during photosynthesis and give off oxygen as a byproduct of the process. Animals—including you—take in oxygen during cellular respiration and give off carbon dioxide as a byproduct of the process.

OBJECTIVES

Detect the presence of carbon dioxide in your breath before and after exercise.

Compare the data for carbon dioxide in your breath with the data from your classmates.

MATERIALS

- eyedropper
- flask, Erlenmeyer, 250 mL (2)
- gloves, protective
- graduated cylinder, 100 mL
- paper towels
- phenol red indicator solution
- stopwatch, or clock with a second hand
- straw, plastic drinking (2)
- water, 200 mL

SAFETY INFORMATION 

PROCEDURE

1. Find a partner. Put on your gloves, safety goggles, and apron.

2. Use the graduated cylinder to pour 100 mL of water into a 250 mL flask.

3. Repeat step 2 so that you have two flasks, each containing 100 mL of water.

4. Using an eyedropper, carefully place four drops of phenol red indicator solution into the water of the first flask. The water should turn orange.

5. Place a plastic drinking straw into the solution of phenol red and water. Drape a paper towel over the flask to prevent splashing.

6. Make sure your lab partner has the stopwatch ready. You will soon exhale into the straw. Your lab partner should begin keeping time as soon as you start to exhale through the straw.

7. Breathe normally, but exhale into the straw in the solution. **Caution:** Do not inhale through the straw. Do not drink the solution, and do not share a straw with anyone. Have your lab partner time how long the solution takes to change color. Record the time.

____

____

Carbon Dioxide in Respiration *continued*

8. Do jumping jacks or sit-ups for 3 min, and repeat steps 4–7 using the second flask.

ANALYZE THE RESULTS

9. Describing Events Describe what happened to the indicator solution.

____

____

10. Organizing Data On a separate sheet of paper, make a table to record your data. Collect data from your classmates and add this to your table.

11. Examining Data Look at your table. What was the longest length of time it took to see a color change without exercise? What was the shortest? How do you account for the difference? Repeat for time with exercise.

____

____

____

12. Examining Data Describe any differences in the data in columns two and three. What could be the cause of these differences?

____

____

13. Constructing Graphs Make a bar graph that compares your data for time without exercise with the data of your classmates.

Carbon Dioxide in Respiration *continued*

DRAW CONCLUSIONS

14. Interpreting Information Do you think that there is a relationship between the length of time the solution takes to change color and the person's physical characteristics, such as gender or athleticism? Explain your answer.

____

____

____

____

____

BIG IDEA QUESTION

15. Making Predictions Describe the organ systems that contribute to getting rid of carbon dioxide, and what would happen if they failed to remove excess carbon dioxide from the body.

____

____

____

____

 DATASHEET

Evaluating Resources for Research

Teacher Notes

In this activity, students will gain hands-on experience finding and selecting research materials from print and electronic sources (covers standard 7.7.b). After completing this activity, students should be able to select reliable and current source material for future research projects.

Evaluating Resources for Research

INVESTIGATION AND EXPERIMENTATION

7.7.b Use a variety of print and electronic resources (including the World Wide Web) to collect information and evidence as part of a research project.

TUTORIAL

Before writing a paper, you must gather information. This process can leave you with stacks of books and thousands of Internet sites. Reading all of these sources would take a very long time. How do you know which sources will have the most relevant information? The procedure below can help you determine the best sources for your research.

Procedure

1. Select the keywords of your topic. Keywords should be specific.

2. Create a research map similar to the one below.

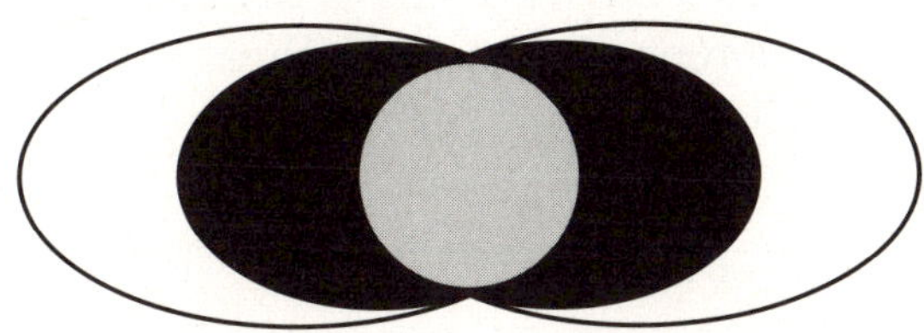

3. Put your keywords in the center circle. Think of broader topics, and place those in the areas outside the center circle. The example below shows a research map on the topic of whale sharks.

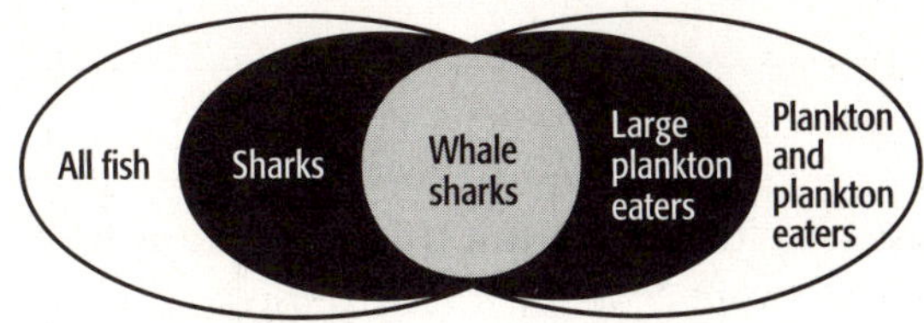

4. List the sources where you may find information about each of your topics including books, CD-ROMs, science magazines, and Web sites.

5. List places where you may find these sources, including computers and libraries.

6. Begin your search with the sources that seem most relevant or have the most information.

Evaluating Resources for Research *continued*

7. Scan the books or online articles. Check or print out sources that have good information. Read what you have gathered.

YOU TRY IT!

Collect information for a research project on the topic of emphysema. Emphysema is a respiratory disorder that happens when alveoli in the lungs have been damaged. Emphysema is often associated with smoking cigarettes.

Procedure

1. Constructing Maps On a separate sheet of paper, create a research map to identify your keywords and the broader topics that may be helpful when finding sources.

2. Organizing Data Use the topics on your research map to find at least eight sources of information. List your sources by type. For example, all of the textbook sources can be listed together.

Analysis

3. Evaluating Data Which of your sources have the most relevant information? Choose the three best sources. Explain why these sources are the most relevant to your research.

____

____

____

4. Identifying Patterns You may find sources with information that is out of date or unscientific. These sources should be dismissed. How will you determine which sources are legitimate and which are not?

____

____

____

5. Evaluating Methods Why do you think it is important to map specific keywords and broader topics? What might happen if you researched only specific topics? What might happen if you researched only broad topics?

____

____

____

Answer Key

Directed Reading A

SECTION: THE CARDIOVASCULAR SYSTEM

1. B	13. D
2. A	14. E
3. D	15. C
4. C	16. B
5. D	17. A
6. A	18. B
7. B	19. D
8. D	20. D
9. B	21. A
10. B	22. C
11. B	23. B
12. A	24. C

SECTION: BLOOD

1. A	14. C
2. D	15. B
3. C	16. C
4. C	17. B
5. A	18. A
6. A	19. A
7. B	20. B
8. B	21. C
9. D	22. C
10. A	23. D
11. B	24. B
12. A	25. A
13. A	

SECTION: THE RESPIRATORY SYSTEM

1. A	11. A
2. A	12. B
3. C	13. A
4. B	14. B
5. B	15. B
6. C	16. A
7. A	17. B
8. D	18. C
9. B	19. C
10. D	

Directed Reading B

SECTION: THE CARDIOVASCULAR SYSTEM

1. heart
2. blood vessel
3. blood vessels
4. cardiovascular system
5. Answers may vary. Sample answer: The cardiovascular system helps maintain homeostasis by carrying nutrients to your cells, removing wastes from your cells, and by carrying hormones throughout the body.
6. B
7. A
8. B
9. B
10. A
11. C
12. to prevent blood from flowing backward
13. the valves
14. arteries
15. capillaries
16. veins
17. pulse
18. blood vessels
19. a capillary
20. veins
21. arteries
22. C
23. D
24. C
25. pulmonary
26. systemic
27. body
28. exercise
29. heart diseases
30. D
31. F
32. E
33. C
34. heart attack

SECTION: BLOOD

1. 5
2. red blood cells, white blood cells, platelets, plasma
3. plasma
4. carry oxygen and nutrients through the body
5. oxygen you inhale
6. platelets
7. destroy pathogens and help clean wounds

8. viruses
9. antibodies
10. white blood cells
11. temperature
12. enlarge
13. B
14. when the ventricles contract
15. when the ventricles relax
16. heart and kidney
17. D
18. B
19. They attach to type B RBCs.
20. Antibodies may react and cause the blood to clump.
21. D
22. lost blood
23. hemophelia, leukemia

SECTION: THE RESPIRATORY SYSTEM

1. energy
2. A
3. D
4. B
5. C
6. H
7. G
8. F
9. E
10. B
11. A
12. C
13. diaphragm, rib muscles
14. diaphragm
15. rib muscles
16. oxygen
17. red blood cells
18. energy
19. carbon dioxide
20. B
21. B
22. carbon dioxide

Vocabulary and Section Summary A

SECTION: THE CARDIOVASCULAR SYSTEM

1. cardiovascular system: a collection of organs that transport blood throughout the body
2. artery: a blood vessel that carries blood away from the heart to the body's organs
3. capillary: a tiny blood vessel that allows an exchange between blood and cells in tissue
4. vein: in biology, a vessel that carries blood to the heart
5. pulmonary circulation: the flow of blood from the heart to the lungs and back to the heart through the pulmonary arteries, capillaries, and veins
6. systemic circulation: the flow of blood from the heart to all parts of the body and back to the heart

SECTION: BLOOD

1. blood: the fluid that carries gases, nutrients, and wastes through the body and that is made up of platelets, white blood cells, red blood cells, and plasma
2. blood pressure: the force that blood exerts on the walls of the arteries

SECTION: THE RESPIRATORY SYSTEM

1. respiration: in biology, the exchange of oxygen and carbon dioxide between living cells and their environment; includes breathing and cellular respiration
2. respiratory system: a collection of organs whose primary function is to take in oxygen and expel carbon dioxide
3. pharynx: the passage from the mouth to the larynx and esophagus
4. larynx: the area of the throat that contains the vocal cords and produces vocal sounds
5. trachea: the tube that connects the larynx to the lungs
6. bronchus: one of the two tubes that connects the lungs with the trachea
7. alveolus: any of the tiny air sacs of the lungs where oxygen and carbon dioxide are exchanged

Vocabulary and Section Summary B

THE CARDIOVASCULAR SYSTEM

1. cardiovascular
2. pulmonary
3. capillaries
4. systemic
5. arteries
6. veins

G	L	Y	M	X	A	T	I	C	I	M	E	T	S	Y	S	A	M
T	I	A	L	V	E	E	L	I	T	O	L	N	M	O	V	L	O
O	B	S	U	R	D	S	H	U	O	C	T	E	F	O	S	X	S
N	E	L	A	O	E	C	P	A	B	C	A	C	L	Q	Y	M	C
S	H	T	G	I	N	T	H	Y	T	A	P	L	I	M	R	P	A
I	I	L	R	O	X	Y	L	N	D	R	X	N	Y	R	A	H	P
L	B	A	R	O	D	D	T	B	I	D	I	T	R	E	N	N	I
S	I	B	L	B	I	P	C	Y	G	I	S	A	O	S	O	O	L
E	B	M	A	R	A	L	R	S	U	O	B	O	T	P	M	G	L
L	M	L	H	R	P	A	H	E	E	V	I	X	A	I	L	E	A
C	A	H	G	J	H	T	U	I	S	A	R	L	P	K	U	S	R
I	W	R	O	R	R	E	M	R	P	S	C	S	D	T	P	U	I
R	X	N	H	R	A	L	E	E	L	C	U	B	P	R	S	I	E
T	C	T	V	S	Y	E	Y	T	E	U	P	R	S	I	E	U	S
T	C	A	N	D	M	T	K	R	E	L	I	E	E	O	S	S	R
E	P	E	M	L	A	S	M	A	N	A	Y	M	L	N	T	E	I
V	E	I	N	S	U	Z	Y	H	T	R	A	Z	H	E	A	S	N

SECTION: BLOOD

Across

2. blood pressure
5. hemophelia
7. leukemia
8. blood
11. white blood cells
14. antibodies

Down

1. systolic
3. pathogens
4. red blood cells
5. hemoglobin
6. antigens
9. shock
10. platelets
12. transfusion
13. diastolic

SECTION: THE RESPIRATORY SYSTEM

1. respiration
2. pharynx
3. trachea
4. larynx
5. bronchi
6. respiratory system
7. alveoli
8. takes in oxygen and releases carbon dioxide

Reinforcement

MATCHMAKER, MATCHMAKER

Blood	O?	A?	B?	AB?
O	yes	no	no	no
A	yes	yes	no	no
B	yes	no	yes	no
AB	yes	yes	yes	yes

1. Answers may vary. Sample answer: I think a hospital would find type O blood the most useful because it can be given to anyone.

Critical Thinking

1. Answers may vary. Sample answer: Normal red blood cells are round and flexible. They contain hemoglobin molecules, which carry oxygen. Sickle cells are sticky and rigid, with abnormal hemoglobin molecules.
2. Answers may vary. Sample answer: Rigid, sticky cells could cause blood clots. Abnormal hemoglobin may carry oxygen poorly. These problems can decrease the oxygen supply to tissues, leading to severe pain.
3. Answers may vary. Sample answer: Because of their rigid structure, more sickle cells are destroyed in the spleen than normal cells are. Mr. Smith has needed transfusions to replace his broken sickle cells with normal red blood cells.
4. Students will likely agree with the doctor. Make sure all answers are well supported by evidence. Answers may vary. Sample answer: Yes, the severe pain, the stroke, the blood clots, and the shortness of breath can be explained by a lack of oxygen to body tissues. The frequent blood transfusions can be explained by a decreased number of red blood cells in the body.
5. Answers may vary. Sample answer: Yes, blood transfusions are not a permanent solution, because the patient still produces unhealthy red blood cells in the bone marrow. If successful, bone marrow transplants allow patients to produce their own healthy red blood cells. Frequent transfusions would no longer be needed.

SciLinks Activity

1. Answers may vary. Sample answer: Red blood cells are used in patients who require increased oxygen-carrying capacity in the blood. Plasma is transfused to treat bleeding disorders in which clotting factors are missing.

2. Answers may vary. Sample answer: All blood donations are tested for type (ABO and Rh). This information is critical to ensuring that compatible blood is given to patients in transfusion.

3. Answers may vary. Sample answer: Autologous transfusions are those in which the blood donor and recipient are the same. Allogeneic transfusions are those in which the recipient is someone other than the donor. One reason an allogeneic transfusion would be used is in the case of treating a patient for a blood disorder in which the patient's unhealthy blood must be treated with healthy blood donated by another individual.

Section Review

SECTION: THE CARDIOVASCULAR SYSTEM

1. left atrium → left ventricle → arteries → capillaries in body → veins → right atrium → right ventricle → arteries → capillaries in lungs → veins → left atrium

2. Sample answer: The heart pumps blood to all parts of the body. Blood carries gases and nutrients to cells and carries wastes away from cells. Arteries carry blood away from the heart. Capillaries allow the exchange of materials between cells and blood. Veins carry blood back to the heart.

3. Sample answer: A heart attack happens when heart muscle cells get insufficient oxygen, and the heart muscle is damaged. The heart may stop. Heart failure happens when the heart cannot pump enough blood to meet the body's needs.

4. Sample answer: Valves keep blood from flowing backward.

5. Sample answer: Capillaries are small and have thin walls, so they are near every cell in the body and allow the exchange of materials between cells and blood.

6. Sample answer: A doctor might prescribe aspirin because it keeps substances in blood from sticking together and blocking vessels, which can cause another heart attack.

7. The wrist and the neck are places where arteries are close to the surface of the skin, so the pulse can be felt there.

8. Sample answer: Each heart contraction pumps blood into arteries, which increases pressure in the arteries. This pressure is the systolic blood pressure.

9. 2,800,000 L/y ÷ (70 beats/min × 60 min/h × 24 h/d × 365 d/y) = 0.076 L/beat = 76 mL/beat

10. Sample answer: A patient may need cardiac bypass surgery if blood vessels near the heart were blocked from atherosclerosis. Such a blockage increases the risk of heart attack.

SECTION: BLOOD

1. Sample answer: Blood carries oxygen to all cells. Blood pressure is the pressure of blood on arterial walls.

2. B and AB

3. Sample answer: Plasma contains water, minerals, nutrients, sugars, proteins, and other substances that are needed by the cells. Red blood cells carry oxygen to cells in the body. Platelets are cell fragments that clump together to form clots. White blood cells destroy pathogens and dead or damaged cells.

4. Sample answer: A doctor must know a patient's blood type so that the doctor can order the right type of blood if the patient needs a transfusion.

5. Sample answer: When the heart contracts, it squeezes blood out of the heart. Blood pressure is the pressure of blood on arterial walls when the heart is contracted (systolic blood pressure) and relaxed (diastolic blood pressure).

6. Sample answer: As blood circulates in the body, it picks up heat from the cells and carries it through blood vessels that are close to the surface of the skin. Close to the skin's surface, the heat leaves the blood and is released into one's surroundings. Blood vessels dilate when the body is hot, which brings blood closer to skin's surface, and constrict when the body is cold, which keeps blood from the surface.

7. Sample answer: A person whose red blood cells cannot deliver oxygen to the body's cells may often feel tired or weak. In addition, the lack of oxygen may damage some of the person's cells, tissues, or organs.

8. Sample answer: The function of blood shown in the image is defense against disease. White blood cells attack and destroy disease-causing pathogens.

9. $174 \div 120 \times 100 = 145\%$ of normal

SECTION: THE RESPIRATORY SYSTEM

1. SARS is caused by a virus, emphysema is caused by damaged alveoli, and some asthma attacks are caused by irritants such as dust or pollen.

2. Sample answer: Air is sucked into or forced out of the lungs. The diaphragm contracts and increases chest-cavity volume, which creates a vacuum. The vacuum pulls air in through the nose or mouth. The air then travels through the pharynx, larynx, trachea, and bronchi to reach the lungs.

3. Sample answer: The respiratory system brings in oxygen, which is carried by the cardiovascular system to cells in the body. The cardiovascular system brings carbon dioxide to the lungs so that it may be exhaled.

4. Sample answer: Fluid in the lungs would prevent the exchange of gases. So, a person would not get all of the oxygen needed to maintain normal activity. The person may feel sick or tired, or die.

5. Sample answer: The statistics indicate that not every child who has asthma had an asthma attack last year.

6. $6 \text{ L} - 3.6 \text{ L} = 2.4 \text{ L}$ exhaled; $2.4 \text{ L} \div 06 \text{ L} \times 100 = 40\%$ of total lung capacity cannot be exhaled.

7. Sample answer: Smoking might coat alveoli with harmful byproducts. This may damage the alveoli and cause emphysema. If the alveoli are damaged, the lungs are not able to exchange gases as efficiently, so a person may feel weak or tired.

Chapter Review

1. Sample answer: The word *function* means "a use or purpose."

2. Red blood cells

3. Arteries

4. blood pressure

5. A

6. D

7. C

8. A

9. Sample answer: Pulmonary circulation carries blood from the heart to the lungs, and then back to the heart. Systemic circulation carries blood from the heart to the rest of the body, and then back to the heart.

10. Sample answer: The first number represents systolic pressure, which is the pressure in the arteries when the ventricles contract. The second number is diastolic pressure, which is the pressure in the arteries when the ventricles relax.

11. Sample answer: In the cardiovascular system, the heart is an organ, blood is a tissue, and red blood cells and white blood cells are both types of cells.

12. A, right atrium

13. B, left atrium

14. C, right ventricle

15. Sample answer: Oxygen enters the body through the nose or mouth; travels through the pharynx, larynx, and trachea to the lungs; and enters the alveoli. In the alveoli, oxygen moves through capillary walls into blood. Red blood cells carry the oxygen through the blood vessels of the cardiovascular system to the heart, which pumps the oxygenated blood to all parts of the body.

16. Answers may vary, but should demonstrate an understanding of why people need transfusions.

17. An answer to this exercise can be found at the end of the Teacher Edition.

18. Sample answer: Red blood cells transport oxygen throughout the body. White blood cells fight pathogens, release antibodies, and destroy cells that have died or been damaged.

19. Sample answer: In a very short time, I would probably die because my cells could not get the oxygen that they need to survive.

20. Sample answer: The immune system produces white blood cells to fight disease-causing pathogens. A high white blood cell count may indicate that a person has an infection.

21. Sample answer: The people who walk through a revolving door are like blood cells. Like the heart, the revolving door works only in one direction; blood cells—or people, in the case of the revolving door—cannot go backwards.

22. Sample answer: Heart attacks affect the heart, and strokes affect the brain. Damage to the heart or the brain can result in the death of an organism.

23. Sample answer: A person who has hemophilia bleeds for a much longer period of time than a person who does not have hemophilia. A person who has hemophilia may bleed so much that he or she dies.

24. **a.** 5,000,000 RBCs/mm^3 × 1,000 mm^3/mL = 5 billion RBCs/mL blood
    b. 5,000,000,000 RBCs/mL × 473 mL/pt = 2.365 trillion RBCs/pt blood

25. B +

Section Quizzes

SECTION: THE CARDIOVASCULAR SYSTEM

1. D
2. A
3. C
4. F
5. B
6. E
7. A
8. D
9. A
10. C

SECTION: BLOOD

1. B
2. A
3. C
4. B
5. A
6. D

SECTION: THE RESPIRATORY SYSTEM

1. G
2. E
3. C
4. F
5. B
6. D
7. A
8. C
9. B
10. A

Chapter Test A

1. C
2. A
3. D
4. B
5. A
6. B
7. A
8. D
9. C
10. D
11. D
12. C
13. B
14. A
15. A
16. D
17. B
18. C
19. respiratory system
20 nose
21. larynx
22. bronchi

Chapter Test B

1. C
2. A
3. D
4. B
5. A
6. C
7. B
8. A
9. D
10. D
11. D
12. C
13. D
14. D
15. C

16. A
17. L
18. H
19. E
20. F
21. B
22. D
23. C
24. B
25. A
26. F
27. D
28. G
29. E

Chapter Test C

1. cellular
2. ventricles
3. pulmonary
4. arteries
5. veins
6. capillaries
7. D
8. D
9. B
10. B
11. A
12. C
13. A
14. hemoglobin
15. It allows them to release energy stored in molecules of sugar and glucose.
16. breathing and cellular respiration
17. to circulate blood, gases, and nutrients throughout the body
18. Answers may vary. Sample answer: A stroke is a cardiovascular disorder that results from a rupture of a blood vessel in the brain. This restricts oxygen to that part of the brain, causing brain cells to die. Asthma is a respiratory disorder in which bronchioles narrow and restrict the body's ability to get the amount of oxygen it needs.
19. Oxygen enters through the nose and passes through the pharynx or throat. It then enters the larynx, which connects the throat to the trachea. The trachea then splits into two tubes called bronchi that run into each lung.

20. Answers may vary. Sample answer: If a person with type A blood is given a transfusion of type B blood, his or her type B antibodies would attach themselves to type B red blood cells. This would lead to clumping of the red blood cells, which could block blood vessels and lead to heart attack or stroke.
21. Answers may vary. Sample answer: These systems interact in the complex task of helping the body receive energy, dispose of waste materials, and fight disease.
22. a. blood vessels, **b.** plasma, **c.** red blood cells, **d.** platelets, **e.** antigens, **f.** pathogens

Performance-Based Assessment

3. Blood vessels that carry blood away from the heart are called arteries.
4. Blood vessels that carry blood toward the heart are called veins.
5. Blood in the arteries that lead to the lungs and away from the heart carry oxygen-poor blood; therefore, not all arteries carry oxygen-rich blood.
6. The right side of the heart has oxygen-poor blood flowing through it.
7. The left side of the heart has oxygen-rich blood flowing through it.
8. Oxygen-poor blood leaves the right ventricle of the heart and enters the lungs; oxygen-rich blood leaves the lungs and enters the left atrium of the heart.
9. The left ventricle has the strongest pumping action, because it must pump blood to the whole body.
10. Answers may vary. Sample answer: The cardiovascular system and the respiratory system work together to supply oxygen to the body's cells so they can produce energy. These systems also work together to remove waste products such as carbon dioxide from the body.

Explore Activity

DATASHEET A

3. yes

4. no

5. Sample answer: No, the paper cone worked like a one-way valve. It allowed the marbles to move in only one direction through the tube.

6. Sample answers: When the marbles are put into the wide end of the cone first, the flaps spread apart. This permits movement of the marbles through the valve. When marbles are put in the narrow end of the cone first, the flaps collapse inward and seal the passageway.

7. Sample answer: Blood could flow in the wrong direction.

DATASHEET B

5. Sample answer: The paper cone worked like a one-way valve. It allowed the marbles to move in only one direction through the tube.

6. Sample answer: When the marbles are put into the wide end of the cone first, the flaps spread apart. This permits movement of the marbles through the valve. When the marbles are put in the narrow end of the cone first, the flaps collapse inward and seal the passageway.

7. Sample answer: Within the heart, valves direct the flow of blood, preventing blood from flowing backward. In the veins, valves prevent the blood from ebbing backward.

DATASHEET C

5. Sample answer: The paper cone worked like a one-way valve. It allowed the marbles to move in only one direction through the tube.

6. Sample answer: When marbles encounter the wide end of the cone first, the flaps spread apart. This permits movement of the marbles through the valve. If marbles encounter the narrow end of the cone first, the flaps collapse inward and seal the passageway.

7. Sample answer: Within the heart, valves direct the flow of blood, preventing blood from flowing backward. In the veins, valves prevent the blood from ebbing backward.

Quick Lab: Vessel Blockage

DATASHEET A

3. Sample answer: No. The blockage doesn't allow much air, if any, through the straw.

4. Atherosclerosis is also a blockage. Like wrapping wire around the straw, it prevents blood from moving through a blood vessel.

5. Sample answer: Cells around or past the blockage will not be able to get nutrients, so they will die. This may kill the organism.

DATASHEET B

3. Sample answer: The blockage does not allow much air, if any, through the straws.

4. Atherosclerosis is also a blockage. It prevents blood from moving through a blood vessel.

5. Sample answer: Cells around or past the blockage will not be able to get nutrients, so they will die. This may lead to the death of the organism.

DATASHEET C

3. Sample answer: The blockage does not allow much air, if any, through the straw.

4. Sample answer: Atherosclerosis is also a blockage. It prevents blood from moving through a blood vessel.

5. Sample answer: Cells around or past the blockage will not be able to get nutrients, so they will die. This may lead to the death of the organism.

Quick Lab: Modeling Blood Pressure

DATASHEET A

2. yes

3. yes

4. The balloon represents an artery.

5. Sample answer: The squeezed bulb represents systolic pressure because pressure in the balloon increases, just as a contracted ventricle increases pressure in an artery.

6. 95/60 mm Hg

DATASHEET B

2. The pressure in the balloon increased.

3. The pressure in the balloon decreased.

4. The balloon represents an artery.

5. Sample answer: The squeezed bulb represents systolic pressure because pressure in the balloon increased, just as a contracted ventricle increases pressure in an artery.

6. 95/60 mm Hg

DATASHEET C

2. The pressure in the balloon increased.

3. The pressure in the balloon decreased.

4. The pipet bulb represents the heart, and the balloon represents an artery.

5. Sample answer: The squeezed bulb represents systolic pressure because pressure in the balloon increases just as a contracted ventricle increases pressure in an artery.

6. 95/60 mm Hg

Quick Lab: Replicate Respiration

DATASHEET A

4. yes

5. Sample answers: When the glove at the bottom moves down, a vacuum is created. This is the same thing that happens when the diaphragm moves down. The vacuum causes air to be drawn into the balloon just as the vacuum created by the diaphragm causes air to be drawn into the lungs during respiration.

DATASHEET B

4. The balloon expanded and drew in air.

5. Sample answer: When the glove at the bottom moves down, a vacuum is created. This is the same thing that happens when the diaphragm moves down. The vacuum causes air to be drawn into the lungs during respiration just as the balloon drew in air.

DATASHEET C

4. The balloon expanded and drew in air.

5. Sample answer: When the glove at the bottom moves down, a vacuum is created. This is the same thing that happens when the diaphragm moves down. The vacuum causes air to be drawn into the lungs during respiration just as the balloon drew in air.

Chapter Lab

DATASHEET A

9. Yes. The phenol red solution turned yellowish. Bromothymol blue turns green.

10. Specific data will depend on student observations; accept all reasonable responses

11. Accept all reasonable answers. It is typical for the solution to change color faster when the student is breathing faster after exercise, so the times listed for step 8 should be shorter than those listed for step 7.

12. Sample answer: Exercise decreased the amount of time it took for the solution to change color because the body produces more carbon dioxide during and after exercise.

13. Bar graphs may vary according to the students' observations and the data collected.

14. In general, there should be little difference between genders. An athlete at rest will usually take the longest time to generate a color change in the indicator solution. There are exceptions, and all answers will depend on students' observations.

15. Sample answers: The heart, blood vessels, and lungs all play a role in getting rid of carbon dioxide wastes. If the body could not get rid of carbon dioxide, the respiratory system may fail, which may result in the death of the person.

DATASHEET B

9. The phenol red solution turns yellowish. Bromothymol blue turns green.

10. Specific data will depend on student observations; accept all reasonable responses. Students' tables should be set up according to the directions given in the question.

11. Accept all reasonable answers. It is typical for the solution to change color faster when the student is breathing faster after exercise, so the times listed for step 8 should be shorter than those listed for step 7.

12. Sample answer: Exercise decreased the amount of time it took for the solution to change color because the body produces more carbon dioxide during and after exercise.

13. Bar graphs may vary according to the students' observations and the data collected.

14. In general, there should be little difference between genders. An athlete at rest will usually take the longest time to generate a color change in the indicator solution. There are exceptions, and all answers will depend on students' observations.

15. Sample answer: The heart, blood vessels, and lungs all play a role in getting rid of carbon dioxide wastes. If the body could not get rid of carbon dioxide, the respiratory system would fail, which may result in death.

DATASHEET C

9. The phenol red solution turns yellowish. Bromothymol blue turns green.

10. Specific data will depend on student observations; accept all reasonable responses.

11. Accept all reasonable answers. It is typical for the solution to change color faster when the student is breathing faster after exercise, so the times listed for step 8 should be shorter than those listed for step 7.

12. Sample answer: Exercise decreased the amount of time it took for the solution to change color because the body produces more carbon dioxide during and after exercise.

13. Bar graphs will vary according to the students' observations and the data collected.

14. In general, there should be little difference between genders. An athlete at rest will usually take the longest time to generate a color change in the indicator solution. There are exceptions, and all answers will depend on students' observations.

15. Sample answer: The cardiovascular system and the respiratory system play a role in getting rid of carbon dioxide wastes. If the body could not get rid of carbon dioxide, the result would be death.

Science Skills Activity

DATASHEET

1. Students' research maps should contain the term emphysema as the keyword, as well as suitable broader topics.

2. Accept all reasonable sources of information.

3. Sample answer: Recent scientific sources, such as up-to-date science textbooks and journals, are the most relevant to my search.

4. Sample answer: To determine which sources are legitimate and which are not, I will assess if a source is out of date (e.g. includes information that has since been disproven). I will also note whether my sources include personal Web sites and research funded by an organization that benefits from the results. These sources are not reliable and should be dismissed.

5. Sample answer: Specific topics may yield results that are too specialized to be applied to broader ideas. Researching broad topics may yield results that are not specific enough.